Food in the 21st Century: from Science to Sustainable Agriculture

Mahendra Shah

Maurice Strong

ISBN 0-8213-4757-8

Library of Congress Cataloging-in-Publication data has been applied for.

A C K N O W L E D G M E N T S

In writing this book, our primary objective has been to highlight the unique role and accomplishments of the Consultative Group on International Agricultural Research and to contribute to mobilizing resources for agricultural research to address the challenges of food security, poverty eradication, and sustainable natural resource management in the developing countries.

We wish to thank our colleagues of the CGIAR Third System Review Panel, whose report and advice contributed immeasurably to this book: Bruce Alberts, Kenzo Hemmi, Yolanada Kakabadse, Klaus Leisinger, Whitney MacMillan, Bongiwe Njobe-Mbuli, Emil Salim, and M.S.Swaminathan.

We owe a deep debt of gratitude to Mary Hager who worked closely with us in drafting several key chapters. Her outstanding writing, attention to detail, and thoughtful comments were tremendously helpful.

The authors wish to acknowledge the many insightful and constructive comments and criticisms of CGIAR Center Directors General: Shawki Barghouti, Stein Bie, Lukas Brader, Ronald Cantrell, Adel El-Beltagy, Hank Fitzhugh, Geoffrey Hawtin, Kanayo Nwanze, Per Pinstrup-Anderson, Tim Reeves, Pedro Sanchez, Jeffrey Sayer, Grant Scobie, David Seckler, Meryl Williams, and Hubert Zandstra.

We express our appreciation to a number of colleagues who provided detailed critiques and substantive suggestions: Paramjit Sachdeva, Dennis Avery, Jock Anderson, Shirley Geer, Selcuk Ozgediz, Philip Pardey, Ravi Tadvalkar, and Donald Winkelmann. We owe a special thanks to David Kinley who shepherded the book through the writing and production phases.

The encouragement and guidance of Ismail Serageldin and Alexander von der Osten were invaluable. The CGIAR made available critical information and provided logistical support throughout the entire process.

We appreciate the contribution from many corners of the World Bank and are especially grateful to James D. Wolfensohn for originally recommending a "popular" version of the CGIAR System Review.

The views expressed in this book are solely those of the authors and should not be ascribed to the persons mentioned above or to the CGIAR or the World Bank. Needless to say we remain solely responsible for any errors and omissions.

October, 1999

CONTENTS

F O R E W O R D

*A*s the twenty-first century dawns, some 840 million people go hungry every day and at least 1.3 billion live on less then US$1 per day. If the world can't make progress against hunger and poverty, by year 2025, there could be 4 billion people living on less then US$2 per day and more than 2 billion living in extreme poverty.

Agriculture and rural development are central to the World Bank's poverty reduction mission. One key challenge ahead is to integrate scientific research into agriculture and rural development programs and services including education and health, micro-enterprise financing, agro-processing, and market access.

The Consultative Group on International Agricultural Research (CGIAR) is one of the most successful partnerships in the history of development in terms of scientific advances, training and capacity building, and agricultural development. Over the past three decades, the CGIAR has established an impressive record of accomplishments in all regions of the developing world, contributing greatly to global food production. It is the only international consortium that exists solely to conduct agricultural research that has a direct bearing on poverty and food security in developing countries.

The CGIAR understands that it has a continuing duty to ensure that scientific research serves the cause of the poor and the hungry. At a time when science is moving at a spectacular pace, it is essential that the new and exciting possibilities for increased agricultural productivity and sustainable natural resources management are realized in the developing countries.

The World Bank, a founder, cosponsor, and provider of both intellectual and financial leadership to the CGIAR, is proud of what the Group has so far achieved, and stands ready to support it as it confronts the future challenges.

I very much hope that this book will convince a wider audience, from around the world, including civil society and the private sector, to address this key challenge of harnessing the best of science to combat poverty, hunger, and environmental degradation in the new millennium.

— JAMES D. WOLFENSOHN, PRESIDENT, THE WORLD BANK

CHAIRMAN'S MESSAGE

*T*he CGIAR takes particular pride in the program of independent external reviews through which the effectiveness of the 16 international agricultural research centers it supports is assessed and maintained. The relevance and quality of CGIAR-supported research as well as its impact on development are thus continuously ensured. Periodically, the same principle is extended to the CGIAR system as a whole. An international panel assesses the state of the system and suggests changes or adjustments for the future.

The third system review of the CGIAR was conducted by a distinguished panel – a powerhouse of intellect and a bastion of credibility – chaired by Maurice Strong.

In its report published in October 1998, the panel gave the CGIAR a powerful endorsement, asserting that "there can be no long-term agenda for eradicating poverty, ending hunger, and ensuring sustainable food security without the CGIAR." At the same time, the panel advised the CGIAR to build on past achievements and learn from past weaknesses.

The response of the CGIAR to that endorsement and advice has been to seize the array of opportunities developing in the world of science; to be the driving force in a new international order of partnership; and to rededicate itself to the tasks of helping to rescue and strengthen diminished human lives. Through this process of reinvigoration, we continue to use cutting-edge science as a means of serving the world's disadvantaged and disconnected.

This book, based on the report of the third system review, provides the context of the challenge and of the CGIAR response. It is an invaluable contribution to greater public understanding of the strengths and role of the CGIAR as bridge builder, knowledge broker, and catalyst of scientific cooperation.

– ISMAIL SERAGELDIN, CHAIRMAN, CGIAR

E X E C U T I V E   S U M M A R Y

THE CONSULTATIVE GROUP ON INTERNATIONAL AGRICULTURAL RESEARCH (CGIAR) is a unique global partnership with a compelling international agenda. Established in the early 1970s, the CGIAR works to promote food security, poverty eradication, and the sound management of natural resources throughout the developing world. It is the largest scientific network of its kind.

CGIAR pursues these objectives through the multi-faceted activities of 16 international research centers located around the world. CGIAR members – 58 industrial and developing countries, private foundations, and regional and international organizations – provide vital financial assistance, technical support, and strategic direction. A host of other public and private organizations work with the partnership as donors, research associates, and advisors.

The New Challenges

Almost three decades ago, the world faced a global food shortage that experts predicted would lead to catastrophic famines. That danger was averted because a group of public and private development agencies created a network of international agricultural research centers and a unique alliance, CGIAR, to support the centers. In what came to be known as the Green Revolution, CGIAR scientists found ways to increase the yields of some of the world's most important food crops, and the world's farmers put the innovations to use.

As the new millennium begins, the world faces another food crisis that is just as dangerous – but much more complex – than the one it confronted thirty years ago. Each year the global population climbs by an estimated 90 million people. This means, at the very least, the world's farmers will have to increase food production by more than 50 percent to feed some two billion more people by 2020.

But the numbers don't tell the full story. The challenge confronting the world is far more intricate than simply producing more food, because global conditions are very different than they were on the eve of the Green Revolution. To prevent a crisis, the world community must confront the issues of poverty, food insecurity, environmental degradation, and erosion of genetic resources.

FOOD SECURITY. Feeding the world in the 21st century will require not only food availability, but food security – access to the food required for a healthy and productive life. It means the ability to grow and to purchase food as needed. It also means that people do not have to rely only on staples such as wheat, rice, potatoes and cassava. Food security focuses attention on areas such as income, markets, and natural resources.

The basic statistics on food security are grim. In addition to the expected population growth, the Food and Agricultural Organization of the United Nations (FAO) estimates as many as 840 million people – a number that exceeds the combined populations of Europe, the United States, Canada, and Japan – currently do not have enough to eat. The companion problem of "hidden hunger" – deficiencies of vital micronutrients – affects even more people in the developing world. The shift away from the traditional food staples will make this challenge even more difficult. Simply increasing productivity of wheat and rice alone may not have the impact it did 30 years ago.

POVERTY. Throughout the developing world, poverty is linked to hunger. For example, in sub-Saharan Africa, plagued by malnutrition, every other person is poor. Rural poverty and accompanying malnutrition are usually tied to the small size or poor quality of farm land and limited off-farm incomes. In addition, more women than men live in poverty in the developing world.

THE ENVIRONMENT. Thirty years ago, the Green Revolution's high-yield food crops were the critical factor in preventing global famine. But an increase in crop lands and the extensive use of fertilizer and irrigation were also instrumental. More intensive fishing and the exploitation of new fishing grounds also helped feed the world. As the 20th century draws to a close, environmental concerns rule out using this mix of strategies, which worked in the past, to meet the food and agriculture crisis that looms ahead.

GENETIC RESOURCES. Environmentalists warn that as much as half of the world's remaining 2.5 billion hectares of tropical forest will come under pressure for agricultural expansion as the demand for food grows. The loss of forests would mean more than the loss of trees and the wood, fuel and other products they provide. Disappearing forests threaten the world's gene pool storehouse.

Meeting these new challenges has been made even more difficult because so few opinion leaders are aware that the world faces these urgent food and agriculture problems. This lack of concern is reflected in the fact that public spending for agricultural research has declined sharply over the past three decades, dropping by about 30 percent in the United States, in real terms, since the 1960s.

AN INTEGRATED APPROACH. Given the complex and interlinked components of the overall challenge of feeding the world in the 21st century, it is clear that solutions that deal only with one part – with crop productivity, for instance, or land use, water conservation, and forest protection – will not be sufficient. The issues are connected and must be dealt with as an interlocking, holistic system.

The CGIAR System in Action

The CGIAR system has the combination of resources and integrated approach needed to meet these complex aspects of the looming global crisis in food and agriculture. In fact, members of a distinguished international panel recently concluded that the CGIAR is the only authoritative international scientific organization capable of ensuring that the tremendous capacities of science are made available to address the problems of the poor in the developing world. CGIAR's assets include a nimble structure, an unmatched mix of knowledge, skills, experience, and perspectives, as well as the ability to link scientists, farmers and environmentalists throughout the world. CGIAR's record of accomplishment and willingness to adapt itself to face new challenges began with the Green Revolution and has continued ever since.

The CGIAR research agenda focuses on five principal challenges: increasing productivity, protecting the environment, preserving biodiversity, improving policies, and strengthening national research.

A RECORD OF ACCOMPLISHMENT. From its beginning, CGIAR's scientists have received world-wide acclaim for their accomplishments. Even more impressive than the accolades, though, has been the global spread of the fruits of CGIAR research.

- More than 300 CGIAR-developed varieties of wheat and rice are in use by the world's farmers.
- The Green Revolution doubled productivity of such staples as wheat and rice.
- In India, wheat production on existing acreage nearly tripled, achieving self-sufficiency.
- More than 200 new CGIAR-developed varieties of maize are being grown in 41 countries.
- CGIAR work has produced improved varieties of legumes, roots and tubers, pasture crops, and other cereals.
- CGIAR's research has improved farming techniques and strategies for managing livestock disease, assessing fish stocks, protecting genetic resources, and effectively managing natural resources.
- Some 50,000 researchers and scientists have worked and trained at CGIAR centers.
- More than 600,000 accessions of germplasm are held in CGIAR genebanks.

BANG FOR THE BUCK. Moreover, CGIAR research on wheat, rice, other staple crops, drylands, and resource management has generated a tremendous impact for a relatively small investment. While the CGIAR accounts for less than 2 percent of the total global public sector investment in agricultural research, the work done by CGIAR centers and their partners has catalyzed agricultural research around the world. In all, CGIAR scientists conduct research on some 30 food commodities, dozens of ecosystems, forests, and water.

Putting Science to Work for Change

The CGIAR is also uniquely positioned to make use of three powerful new scientific breakthroughs that have the potential to radically reshape the world's agricultural and food production: the gene revolution, natural resource management, and the information revolution. These striking advances could make the emerging challenge of feeding a growing population in the 21st century much more manageable.

BIOTECHNOLOGY. CGIAR has played a lead role in the development of biotechnology and genetic research for the poor. It has also been integral in weighing the arguments that these breakthroughs have generated. In 1998, the overall CGIAR investment in biotechnology reached US$30 million, accounting for about 7 percent of the total budget. Moreover, the work at the CGIAR centers focuses on biotechnology appropriate to the needs of the poor.

NATURAL RESOURCE MANAGEMENT. However, the CGIAR believes that any gains made through combinations of biotechnology and traditional breeding must

*The CGIAR is now poised to
redouble its efforts to ensure
that our effectiveness in the future
will make a difference between
despair and hope
for the millions now unreached
by the dazzling scientific
advances of our time.*

— Ismail Serageldin, CGIAR Chairman

be fully integrated with sound strategies for managing natural resources. That is why the CGIAR's crop improvement approach emphasizes the vital role natural resource management must play in achieving the goals of sustainable agriculture.

THE INFORMATION REVOLUTION. The information and communications revolution presents a tremendous new opportunity for the CGIAR to bring scientific knowledge and indigenous and local knowledge together to bear on global challenges, and to make that information available to its constituents. The CGIAR is committed to being at the forefront of harnessing these technologies to pursue its mission.

An Action Agenda for the 21st Century

The CGIAR has the system in place and the tools at hand to meet the complex challenges of the looming global crisis in food and agriculture. And it has something else — an action agenda for the 21st century.

THIRD SYSTEM REVIEW PANEL. Since it was formed, the CGIAR system has undergone three extensive reviews by panels of outside experts. Each time, the panels have highlighted the important contributions made by the CGIAR. And each time the panels have recommended adjustments to help CGIAR more ably meet changing global food needs and confront environmental realities.

The most recent outside review, completed in the fall of 1998, called for the CGIAR "to contribute to food security and poverty eradication in developing countries through research, partnership, capacity building, and policy support, promoting sustainable agricultural development based on the environmentally sound management of natural resources." CGIAR members enthusiastically endorsed this recommended expansion of the system's mission.

For its part, the CGIAR has taken on the review panel's recommendations as a challenge to rethink roles and redefine strategy, "to build on past achievements but confront past weaknesses as well," reports CGIAR Chairman Ismail Serageldin. That process is ongoing. As this continues, the Chairman assures that "the CGIAR is now poised to redouble its efforts to ensure that our effectiveness in the future will make a difference between despair and hope for the millions now unreached by the scientific dazzling advances of our time."

RECOMMENDATIONS. To make the difference between despair and hope, the panel identified specific steps the system should take in the following areas:

- Financial Support – The Need for Greater Commitment to CGIAR
- The Private Sector – New Cooperation, Uncompromised Objectives
- The Role of Governments – More Emphasis on Policy Research and Reform
- Partnerships – Applying Science to the Problems of the Poor
- Intellectual Property – Protecting the Interests of the Poor
- Genetic Resources – An Integrated Gene Management Initiative
- Natural Resources – More Careful Management
- Information Exchange – A Global Food Security Knowledge System
- National Agricultural Systems – Building Stronger Partnerships
- Food Security in Africa – A Capacity-Building Initiative
- The CGIAR System – Strategies for Better Focus, Integration and Direction

MORE PREDICTABLE AND STABLE SUPPORT.
To these recommendations, the panel added a warning:

The panel emphasized that CGIAR will only be able to do what is needed if it can expand its sources of funding and secure more predictable and stable support. That requires both bold initiatives from the CGIAR and stepped-up commitments from governments, international organizations, national research organizations, NGOs, and the private sector.

The world enters the 21st century on the brink of a new food crisis that is as dangerous, but far more complicated than the threats it faced in the 1960s. Without question the challenges are global and demand the attention and commitment of all peoples and countries. Because its international status protects it from political pressures and regional or national influences, the CGIAR is in a unique position to deal with problems that cut across boundaries and affect entire regions and ecosystems. It has flexibility and autonomy and, through its network of centers, works cooperatively with both developing and industrial countries. It has a 30-year record of accomplishment.

As the Third System Review Panel said, "There can be no long-term agenda for eradicating poverty and ending hunger without the CGIAR." But the CGIAR cannot do the job alone.

PART I

The CGIAR:
Meeting the Challenge

1. NEW CHALLENGES

ALMOST THREE DECADES AGO, the world faced a global food shortage that experts predicted would lead to catastrophic famines. These were averted because a group of public and private development agencies created a network of international agricultural research centers and a unique alliance, the Consultative Group on International Agricultural Research (CGIAR), to support the centers. In what came to be known as the Green Revolution, CGIAR scientists found ways to increase the yields of some of the world's most important food crops, and the world's farmers put the innovations to use.

At the new millennium begins, the world faces another food crisis that is just as dangerous – but much more complex – than the one it confronted 30 years ago. Producing more food will not, by itself, prevent catastrophe this time. Instead, the world community will have to overcome new challenges including poverty, food insecurity, environmental degradation, and genetic resource preservation. Only the CGIAR has the combination of knowledge, skills, and experience to deal with these diverse issues.

Each year the global population climbs by an estimated 90 million people. This means, at the very least, the world's farmers will have to increase food production 50 percent to feed some two billion more people by 2020, and maybe as much as 100 percent if current trends in increased meat consumption – and the accompanying demand for more feed grains – continue. These stark numbers have led some analysts to believe that what is needed is a new and greener revolution to once again increase productivity and boost production.

MORE COMPLEX CHALLENGES. But the numbers don't tell the full story. The challenge confronting the world is far more complex than simply producing more food because global conditions are very different than they were at the time of the Green Revolution. Easy gains in food crop yields seem to have been achieved and, in some regions, yields may have reached their ceilings. The environment has assumed a new importance. Many of the most promising lands are already under cultivation, erosion is taking a growing toll, water shortages loom in many areas, and the majority of the world's fisheries are overexploited. Feeding the world will also require maintaining biodiversity and addressing the intransigent problems of poverty and poor health so prevalent in the developing world.

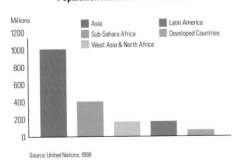

Population Increases: 1995 to 2020

Millions

Asia
Sub-Sahara Africa
West Asia & North Africa
Latin America
Developed Countries

Source: United Nations, 1998

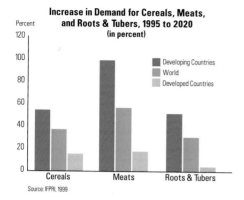

Increase in Demand for Cereals, Meats, and Roots & Tubers, 1995 to 2020
(in percent)

Percent

- Developing Countries
- World
- Developed Countries

Cereals   Meats   Roots & Tubers

Source: IFPRI, 1999

As many as 840 million people –
more than the
combined populations of
Europe, the United States,
Canada, and Japan –
do not have enough to eat.
Among the 40,000 people
who die each day
of malnutrition are
19,000 infants and children.

COMPLACENCY. Meeting these new challenges has been made even more difficult because so few opinion leaders are aware that the world faces urgent food and agriculture problems. The abundance of grain produced in the developed world, and the fact that millions of hungry people in the developing world remain invisible, also obscure the challenges.

This lack of concern is reflected in the fact that public spending for agricultural research has declined sharply over the past three decades. In the United States, as one example, support has dropped by about 30 percent, in real terms, since the 1960s. Official development assistance for agriculture and fisheries in developing countries fell by almost half, from a peak of US$19 billion in 1986, to about US$10 billion just a

**Official Development Assistance
and External Assistance to Agriculture**
(Inflation and Exchange Rate Adjusted)

Index: 1986=100

Official Development Assistance

External Assistance to Agriculture

1986 1987 1988 1989 1990 1991 1992 1993 1994 1995 1996

Source: IFPRI

17

decade later. In the industrialized world, this decline has been masked by the significant increase in agricultural research investment in the private sector. In the developing world, however, there has been little increase in private investment to offset the public sector decline.

Food Security

Feeding the world in the 21st century will require not only food availability, but food security – access to the food required to lead a healthy and productive life. For specific populations, it means the ability to grow and to purchase food as needed. It also means that people do not have to rely only on staples such as wheat, rice, potatoes, and cassava. Food security focuses attention on areas such as income, which must be sufficient to purchase food; markets, which must be competitive to keep prices low; and natural resources, which must be conserved to assure sustainable, long-term productivity.

Number of Food-Insecure People, 1995

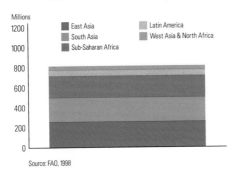

Source: FAO, 1998

The statistics on food security are grim. In addition to the expected population growth, the FAO estimates as many as 840 million people – a number that exceeds the combined populations of Europe, the United States,

Canada, and Japan – currently do not have enough to eat. That means about one-fifth of people living in the developing world do not get enough calories, enough protein, or both. The resulting malnutrition is particularly harmful for children and pregnant women. According to the World Health Organization, an estimated 19,000 infants and children are among the 40,000 people who die each day of problems linked to malnutrition.

Number of Malnourished Children, 1995

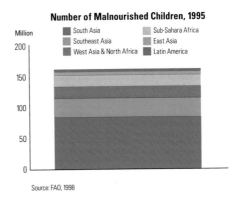

Source: FAO, 1998

HIDDEN HUNGER. The companion problem of "hidden hunger" – deficiencies of vital micronutrients – affects even more people in the developing world. An estimated two billion lack sufficient iron in their diets, with about 1.2 billion weakened by iron deficiency anemia. Vitamin A deficiency affects about 125 million school children and has produced irreversible eye damage in an estimated 14 million. Another 600 million people suffer iodine deficiency disorders.

Hunger of this dimension traditionally has been seen among the rural poor who could not grow enough food to meet their needs. Now it has spread to growing numbers of the urban poor – many of whom have left their rural homes and moved to cities – who are not able to afford the food they need. Often food is available, but not accessible.

CHANGING PATTERNS. Changing consumption patterns in much of the world will make this challenge even more difficult. In much of the world, people are shifting away from the traditional food staples of developing nations. In urban areas, milk, livestock, fish, fresh fruits, and vegetables now either complement or supplant rice and wheat as dietary staples, while in many rural areas, rice and wheat are replacing sorghum, millet, and maize. The result is that simply increasing productivity of wheat and rice may not have the impact it did 30 years ago.

Poverty

Throughout the developing world, poverty is linked to hunger. In sub-Saharan Africa, plagued by malnutrition, every other person is poor. There and in South Asia – another badly undernourished region – more than half of the poor live in rural areas where lands are marginal and their livelihoods depend mainly on agriculture. About one-third of the population of the developing world, an estimated 1.3 billion people, exist on incomes of US$1 a day, or less. Another two billion are only slightly better off, living on the equivalent of US$2 or less per day.

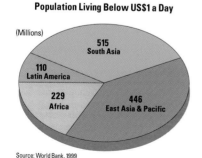

Population Living Below US$1 a Day

(Millions)

515 South Asia

110 Latin America

229 Africa

446 East Asia & Pacific

Source: World Bank, 1999

A Glimpse into the Food Future – Changing Diets in Asia

As Japan's economy grew, per capita rice consumption fell by about 40 percent from the 1960s through the 1990s, while the demand for meat grew by 360 percent. If the rest of Asia follows a similar pattern, the pressure on future food supplies in Asia will be enormous. Already consumption of livestock products in China has grown in response to the new diet patterns of the growing middle class, with poultry consumption predicted to expand by about 7 percent each year, followed by beef at 5.3 percent and pork at 3.1 percent.

To put this in perspective, one CGIAR center, IFPRI, has predicted that the demand for meat and milk in developing countries will more than double over the next 20 years and reported that the gain from the "livestock revolution" over the past three decades was more than double that of the Green Revolution.

ROOT CAUSES. Poverty in the developing world has many roots including political and social discrimination. But many are poor because they have no tangible assets, no land and no livestock, and lack formal education and technical skills. Many either settle where the land is only marginally productive at best and where governments have failed to provide the basic infrastructure essential to economic development, or they migrate to urban areas.

POVERTY AND MALNUTRITION. Rural poverty and accompanying malnutrition are usually tied to land. Much of the land owned by poor families, has been

If we don't do something,

we could have

4 billion people

living on less than US$2 a day

and more than 2 billion

in absolute poverty

in the year 2025.

— World Bank President James Wolfensohn

divided and subdivided so often that what remains cannot support a family. Some of the land is hilly and rocky, subject to erosion and located in regions where water is scarce and soils are not fertile. Many of these areas were largely bypassed by the Green Revolution.

GENDER. Because they rarely can own property or go to school, more women than men live in poverty in the developing world. But as men from poor rural areas have moved to cities to find work to help support their families, the role of the women left behind has changed dramatically. Many have become *de facto*, both farm managers and field laborers. This is especially true in Africa, where much crop cultivation is now done by women.

Environmental Concerns

Thirty years ago, the Green Revolution's high-yield food crops were the critical factor in preventing global famine. But an increase in crop lands and the extensive use of fertilizer and irrigation were also instrumental. Moreover, more intensive fishing and the exploitation of new fishing grounds helped feed the world. As the 20th century draws to a close, environmental concerns rule out using this mix of strategies, which worked in the past, to meet the food and agriculture crisis that looms ahead.

LIMITS ON PRODUCTIVITY. Increases in the productivity of food crops seem to be peaking, even on lands where the Green Revolution was most successful. Current increases in yield have dropped to about 1.3 percent a year, barely half what they were 30 years ago. That drop could signal that irrigation and fertilization have reached their effective limits, or even that the ability of plant breeders to develop higher

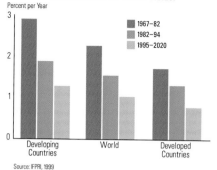

Annual Growth in Cereal Yields:

Percent per Year

Legend: 1967–82, 1982–94, 1995–2020

Categories: Developing Countries, World, Developed Countries

Source: IFPRI, 1999

yielding varieties has reached a plateau, at least for cereal crops.

DEGRADED LAND. The Green Revolution was most effective where soils were fertile and water plentiful. Many of the lands that did not benefit are eroded, infertile, or severely degraded, and located where water is scarce. Such lands will rarely be capable of providing the Green Revolution level of yields, yet the amount of such land continues to grow. Over the past half-century, more than one-quarter of the world's 8.7 billion hectares of agricultural lands, pastures, forest and woodlands have been degraded through misuse or overuse, with another 5 to 10 million hectares added each year.

Many of the most degraded lands are found in the world's poorest countries, in densely populated, rainfed farming areas, where overgrazing, deforestation, and inappropriate use compound problems. When lands become infertile, traditional farmers either let the land lie fallow until it recovers or simply abandon unproductive lands and move on, clearing new forest lands to cultivate. The newly cleared land is then farmed and the process is repeated.

In addition, an estimated 500 million people live on hillsides that are difficult, if not impossible, to till. Another 850 million live in areas threatened by desertification.

WATER SCARCITY. About 70 percent of the world's fresh water supply goes to agriculture, a figure that approaches 90 percent in highly productive Asian countries such as China and India which rely on extensive irrigation. "The shortage of fresh water is looming as the most serious obstacle to food security, poverty reduction and protection of the environment," says CGIAR Chairman Ismail Serageldin, who serves as chairman of the World Commission on Water for the 21st century. "Even if we do everything we can to make irrigated agriculture more water efficient, humanity will still need at least 17 percent more fresh water to meet all of its food needs than is currently available."

That 17 percent shortfall represents what has been called the "world water gap." Failure to bridge this huge gap will lead to higher food prices and expensive

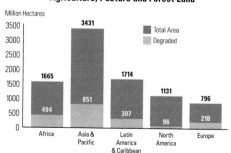

**Regional Soil Degradation
Agriculture, Pasture and Forest Land**

Million Hectares

Legend: Total Area, Degraded

Africa: 1665, 494
Asia & Pacific: 3431, 851
Latin America & Caribbean: 1714, 307
North America: 1131, 96
Europe: 796, 218

Source: IFPRI, 1999

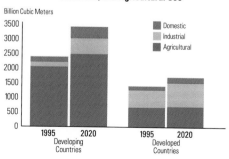

**Water Withdrawals for Domestic,
Industrial, and Agricultural Use**

Billion Cubic Meters

Legend: Domestic, Industrial, Agricultural

1995, 2020 Developing Countries
1995, 2020 Developed Countries

Source: IFPRI, 1999

21

Within 25 years,

one-quarter of humanity

is likely to suffer

severe water scarcity.

This crisis poses

the greatest threat

to food security,

human health,

and the environment.

— International Water Management Institute, Sri Lanka

food imports for water scarce countries – which are many of the same developing countries with serious land degradation problems. Already, the commission reports, 31 countries – almost all in the developing world – face water shortages. By 2050 the number will rise to 55, assuming no significant change in water use patterns.

Though renewable, fresh water is a finite resource, not evenly distributed across countries, regions or even seasons. Two-thirds of the world's population live in areas that receive only one-quarter of the world's annual rainfall, while such sparsely populated areas as the Amazon Basin receive a disproportionate share. Because of extensive upstream use, some of the world's major rivers – the Nile and the Ganges, for instance – barely run into the sea any more.

Even those countries with enough total water suffer regional or seasonal shortages, often on a regular and predictable basis, making it difficult to distribute water more evenly or find new sources. Public concerns about protecting natural resources have made it

Projected Water Scarcity in 2025

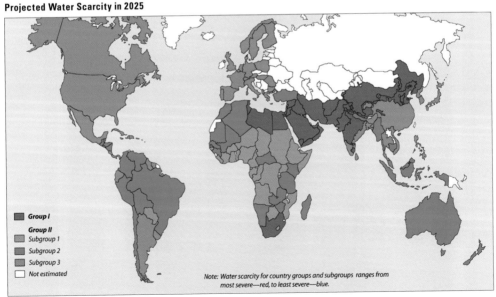

Group I
Group II
Subgroup 1
Subgroup 2
Subgroup 3
Not estimated

Note: Water scarcity for country groups and subgroups ranges from most severe—red, to least severe—blue.

Source: IWMI, 1999

difficult to build new dams, little money is invested in expanding irrigation, and insufficient attention is given to efficient water use, either for agricultural and industrial use or for domestic use. Since water use is often subsidized and consumers rarely pay the full cost, water is a commodity too frequently taken for granted – except where it is scarce.

FISHERIES. More than 70 percent of the world's fishery resources are overexploited and cannot become more productive unless fishing pressures are reduced and stocks allowed to recover. Many of the world's major fishing areas are in serious decline and four of the most productive – the Gulf of Thailand, seas in southeast Asia, the southern part of the North Sea and the northern Mediterranean – have been depleted for commercial purposes. At the same time that stocks are declining, the number of recorded vessels in world fishing fleets has soared from 585,000 in 1970 to more than 3.5 million, not counting untold millions of smaller, unregistered craft.

OTHER PROBLEMS. In several other major environmental problem areas, agriculture seems to be both culprit and victim. Agriculture produces a variety of gases – from animals, from fertilizer, from clearing

Water Disaster in India?

In India, which relies on extensive irrigation, water tables are falling. According to IWMI, India currently withdraws groundwater at twice the recharge rate. Eventually, the shortfall could reduce India's harvests by as much as 25 percent. Since India – well on the way to becoming the world's most populous nation – already maintains an uneasy balance between production and consumption, the impact could be disastrous.

and burning lands – that have been linked to four different problems: increased ozone pollution in the lower atmosphere, acid rain, threatened global warming and a reduction in stratospheric ozone. Agriculture, in turn, may pay the price of these problems. For instance, a study in Pakistan suggests high levels of atmospheric ozone might reduce wheat yields by as much as 30 percent, and forecasters warn that global warming could change climate patterns, temperature and rainfall throughout the world.

The irrigated lands that achieved the greatest gains during the Green Revolution account for less than one-fifth of all land under cultivation, yet they produce nearly half the world's food. Many experts believe those lands will need to produce up to 70 percent if future needs are to be met. Even under the best conditions, production on marginal lands is unlikely to make up the difference. If productive land cannot produce more, additional land will have to be cleared, which means more of the world's already vanishing forests would be lost.

Deforestation

Environmental experts warn that as much as half of the world's remaining 2.5 billion hectares of tropical forest will come under pressure for agricultural expansion as the demand for food grows. The loss of forests would mean more than the loss of trees and the wood, fuel, and other products they provide. Forests protect watersheds, prevent erosion, help minimize the impact of floods and drought, and stabilize local climates. But the main victims will be the nearly 350 million people who depend on the multiple products of the forest for their daily needs.

If forests continue to disappear at the current rate, some biologists warn that as many as 20 percent of all tropical forest species could be extinct within 30 years.

BIODIVERSITY. Perhaps most importantly, disappearing forests threaten the world's gene pool storehouse. Tropical forests harbor at least half of the world's genetic biodiversity – and maybe as much as 90 percent of the estimated 10 million species on earth. In the context of the challenge to CGIAR, maintaining genetic diversity is essential to the future food supply. But, to protect what forests remain and maintain this diversity, the CGIAR must also explore ways to improve production on marginal lands so farmers no longer need to move to new lands.

> *OECD countries now spend more than US$300 billion a year on agriculture subsidies, while the current funding for international centers working on agriculture research targeted for the problems of the poor is less than US$350 million.*

The Role of Policy

The important role of sound public policy decisions can be easily overlooked in the face of the other challenges confronting the world. Yet public policies, especially at the local level, profoundly impact both agricultural development and natural resource management. Decisions in such areas as road-building and transportation infrastructure, land-use planning, and technical choices for public irrigation systems affect how land, water, and other natural resources can be used for agriculture and whether that use will be sustainable.

In addition, natural resources such as water, forests, grazing lands, and wildlife are traditionally managed at the local level. Often these resources are considered common property with little consideration to protecting and preserving them for the future. As pressures increase to expand agricultural production and, at the same time, conserve natural resources through wise use and management, the crucial role of public policy cannot be ignored, nor can the need for public education, which must support the process that leads to effective local policy decision-making.

2. THE CGIAR SYSTEM IN ACTION

THE CGIAR SYSTEM HAS THE COMBINATION OF RESOURCES needed to meet the many complex aspects of the looming global crisis in food and agriculture: food security, poverty, protecting the environment, and genetic resource preservation. In fact, members of a distinguished international panel who recently reviewed the system concluded that the CGIAR is the only authoritative international scientific organization capable of ensuring that the tremendous capacities of science, particularly biotechnology and gene management – integrated with natural resource management – are made available to address the problems of the poor in the developing world. CGIAR's assets include a nimble structure, an unmatched mix of knowledge, skills, experience, and perspectives, as well as the ability to link scientists, farmers, and environmentalists throughout the world. CGIAR's record of accomplishment and willingness to adapt itself to face new challenges began with the Green Revolution and has continued ever since.

The need for a special partnership within the global agricultural research community focused on combating hunger and poverty through productivity-oriented research was first perceived in the late 1960s, in response to the specter of widespread famine in parts of Asia. Internationally-supported work at several research centers to develop new varieties of rice and wheat had begun to generate unprecedented harvests. Their success fueled hopes that a "Green Revolution" of agricultural modernization could be extended worldwide.

THE BIRTH OF CGIAR. Leaders from 18 international organizations, foundations, and concerned governments formally joined together in 1971 as the first Members of the CGIAR. Through their continuing support, hundreds of new wheat and rice varieties were developed, released, and planted during the Green Revolution, adding an estimated US$50 billion to the value of global food supplies over two decades. The food-to-population balance in the most threatened countries was transformed and the threat of widespread famine was averted.

NEW CENTERS. To build on the early achievements, new CGIAR Centers were established to pioneer improvements in other key food crops, such as legumes, roots and tubers, and other cereals, and to concentrate on better management of livestock. Centers were likewise created to evaluate the special problems of dry, semi-arid and tropical regions, and to conduct research on pressing issues involving forestry, agroforestry, water management, and fish and marine resources. Centers were also set up to examine national and international food policies, and to build the capacity of agricultural research at the national level.

THE CGIAR TODAY. Today, CGIAR is largest scientific partnership in this field. The 16 CGIAR Centers around the world are harnessing cutting-edge knowledge to help meet the world's enormous food needs – with a steadfast allegiance to scientific excellence and the public welfare. All of the benefits of CGIAR research are kept within the public domain; new production and natural resource management technologies and plant varieties are available free to interested parties throughout the world.

The 16 CGIAR Centers Around the World Are Harnessing Cutting-edge Knowledge to Help Meet the World's Future Food Needs.

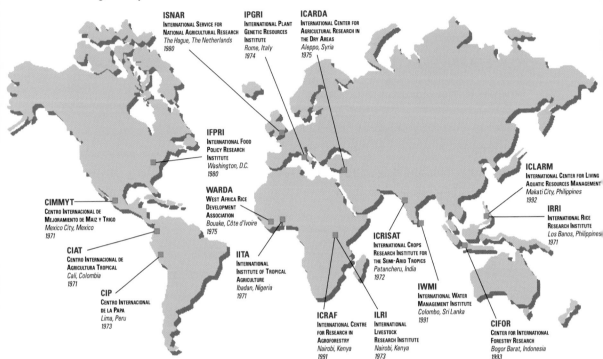

ISNAR
INTERNATIONAL SERVICE FOR
NATIONAL AGRICULTURAL RESEARCH
The Hague, The Netherlands
1980

IPGRI
INTERNATIONAL PLANT
GENETIC RESOURCES
INSTITUTE
Rome, Italy
1974

ICARDA
INTERNATIONAL CENTER FOR
AGRICULTURAL RESEARCH IN
THE DRY AREAS
Aleppo, Syria
1975

IFPRI
INTERNATIONAL FOOD
POLICY RESEARCH
INSTITUTE
Washington, D.C.
1980

ICLARM
INTERNATIONAL CENTER FOR LIVING
AQUATIC RESOURCES MANAGEMENT
Makati City, Philippines
1992

CIMMYT
CENTRO INTERNACIONAL DE
MEJORAMIENTO DE MAIZ Y TRIGO
Mexico City, Mexico
1971

WARDA
WEST AFRICA RICE
DEVELOPMENT
ASSOCIATION
Bouake, Côte d'Ivoire
1975

IRRI
INTERNATIONAL RICE
RESEARCH INSTITUTE
Los Banos, Philippines
1971

CIAT
CENTRO INTERNACIONAL DE
AGRICULTURA TROPICAL
Cali, Colombia
1971

IITA
INTERNATIONAL
INSTITUTE OF TROPICAL
AGRICULTURE
Ibadan, Nigeria
1971

ICRISAT
INTERNATIONAL CROPS
RESEARCH INSTITUTE FOR
THE SEMI-ARID TROPICS
Patancheru, India
1972

CIP
CENTRO INTERNACIONAL
DE LA PAPA
Lima, Peru
1973

IWMI
INTERNATIONAL WATER
MANAGEMENT INSTITUTE
Colombo, Sri Lanka
1991

ICRAF
INTERNATIONAL CENTRE
FOR RESEARCH IN
AGROFORESTRY
Nairobi, Kenya
1991

ILRI
INTERNATIONAL
LIVESTOCK
RESEARCH INSTITUTE
Nairobi, Kenya
1973

CIFOR
CENTER FOR INTERNATIONAL
FORESTRY RESEARCH
Bogor Barat, Indonesia
1993

The CGIAR's research agenda focuses on both strategic and applied research. It extends across the entire range of problems affecting agricultural productivity, and links these to the broader concerns for poverty eradication, sustainability, biodiversity, natural resource management, capacity building, and policy improvement. Today, CGIAR scientists and researchers are working in more than 100 countries.

The CGIAR research agenda focuses on five principal challenges:

INCREASING PRODUCTIVITY. The CGIAR strives to make the agriculture of developing countries more productive through genetic improvements in plants, livestock, fish, and trees and through better management practices.

PROTECTING THE ENVIRONMENT. Conserving natural resources, especially soil and water, and reducing the impact of agriculture on the surrounding environment is an essential and growing part of research efforts. The CGIAR plays a leading role in identifying and promoting sustainable agricultural ecosystems and in developing solutions to pressing environmental problems.

PRESERVING BIODIVERSITY. The CGIAR's collections of plant genetic resources, contain more than 600,000 accessions of more than 3,000 crop, forage, and pasture species. The collections include improved varieties and, in substantial measure, the wild species from which those varieties were created. Duplicates of these materials are freely available to researchers around the world so that new gene combinations can be brought to bear on current problems. The CGIAR was the first organization to place its collections under the auspices of the FAO as the basis of an international network of *ex situ* collections.

IMPROVING POLICIES. Agricultural production and natural resource management are heavily influenced by public policy. The CGIAR's policy research aims to help streamline and improve policies that strongly influence the spread of new technologies and the management and conservation of natural resources.

STRENGTHENING NATIONAL RESEARCH. The CGIAR supports national agricultural research in developing countries through collaborative work with colleagues in national programs, strengthening of skills in research administration and management, and formal training programs for research staff. Such investments have consistently paid handsome dividends.

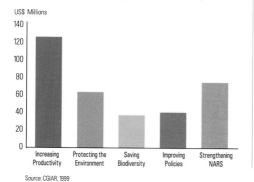

1998 Investments by CGIAR Undertakings

US$ Millions

Source: CGIAR, 1999

A 30-Year Record of Achievement

From its beginning, CGIAR's scientists have received world-wide acclaim for their accomplishments including the Nobel Peace Prize to Norman Borlaug, and the King Baudouin Foundation International Development Prize to the CGIAR system. In recent years, almost half of the World Food Prizes have been awarded to CGIAR researchers.

THE GLOBAL SPREAD OF CGIAR RESEARCH. Even more impressive than the accolades, though, has been the global spread of the fruits of CGIAR research. Moreover, CGIAR research has generated a huge impact for a relatively small investment. While the CGIAR system accounted for less than 2 percent of the global public investment in agricultural research, the work done by CGIAR centers and their partners has catalyzed agricultural research around the world. In all, CGIAR scientists conduct research on some 30 food commodities, dozens of ecosystems, as well as fisheries, livestock, forestry, and water.

At the same time, policy research done by CGIAR's IFPRI identified ways poorly-designed public policies, such as food subsidies, slow economic growth while properly designed environmental and resource protection strategies can move national systems toward sustainability. The support provided by experts at CGIAR's ISNAR has helped more than 60 developing countries improve their agricultural research policies, organizations, and management.

Ultimately, though, the CGIAR collection of germplasm may prove to be one of the system's most significant contributions to future food security.

CGIAR Gene Banks: Safeguarding the World's Food Future

Since agriculture began farmers have always saved seeds and materials from the strongest and best plants for the next crop. Through this process, farmers developed literally thousands of folk varieties, or "landraces," of edible crops.

Not too many generations ago, the once informal process became more scientific as commercial companies and their scientists took over the task of improving crops such as hybrid corn and soybeans. At the same time, scientists at the CGIAR centers were developing higher yielding varieties of rice, wheat, and other staples. As they did, crops tended to become more uniform and the traditional varieties

began to disappear. China, which once had an estimated 10,000 landrace varieties of wheat, now has only about 1,000. No one knows what genetic traits that lead to insect and disease resistance, stronger plants, higher yields, or even better tasting crops, might have been lost.

Within the past half-century scientists began to worry about the loss and established gene banks to preserve what remains. There are now more than six million "accessions" stored in genebanks throughout the world, more than 600,000 of them held in trust for the world community in the genebanks of 11 CGIAR centers under the auspices of the FAO.

Top Three Holders of Major Food Crops (Size of Collection)

Crop	Major Holders		
	Largest	Second largest	Third largest
Banana/Plantain	IPGRI	France	Honduras
Cassava	CIAT	Brazil	IITA
Chickpea	ICRISAT	ICARDA	Pakistan
Cowpea	IITA	Philippines	USA
Faba bean	ICARDA	Germany	Italy
Groundnut	USA	India	ICRISAT
Lentil	ICARDA	USA	Russia
Phaseolus	CIAT	USA	Mexico
Potato	CIP	Colombia	Germany
Rice	IRRI	China	India
Sorghum	ICRISAT	USA	Russia
Sweet potato	CIP	Japan	USA
Wheat	CIMMYT	USA	Russia
Yam	IITA	Côte d'Ivoire	India

Source: FAO, 1996

CGIAR Training in Germplasm Collecting, Conservation and Genebank Operations

Centre	Period	Type of Training (number of trainees)				
		Courses	On-the-job	Theses	Total	Countries
CIAT	1983-95	65	47	49	161	22
CIP	1979-94	484	21	–	505	28
ICARDA	1990-94	99	–	–	99	18
ICRISAT	1976-95	–	337	6	343	–
IITA	1992-94	47	–	–	47	19
ILRI	1990-94	61	47	7	115	–
IPGRI	1974-95	1745	84	256	2085	139
IRRI	1980-95	47	45	–	92	–

Source: SGRP, 1997.

Composition of CGIAR, Government and Private Collections

Type of Material	CGIAR	Governmental	Private
Landraces and old cultivars	59%	12%	9%
Wild species and weedy relatives	14%	4%	6%
Advanced cultivars and breeders' lines	27%	18%	47%
Other/mixed material	–	66%	38%

Source: FAO, 1996

Germplasm Distributed by CGIAR Centers, 1992-1994

Recipient	Number of Samples										
	CIAT	CIMMYT	CIP	ICARDA	ICRISAT	IITA	ILRI	IPGRI	IRRI	WARDA	Total
NARS**	6,286	6,292	13,841	24,652	53,319	7,729	2,056	708	11,225	4,200	130,508
NARS***	4,539	66,767	1,015	12,651	906	2,431	210	373	8,428	–	37,320
CGIAR Centers	33,064	17,474		27,224	73,420	11,695	5,088	36	46,404	3,740	218,415
Private sector	331	760		42	4,314	53	644	–	357	15	6,516
Others	85	–	–	–	5,256	–	–	227	7,644	240	13,452
Total	44,505	31,293	14,856	64,569	137,215	22,178	7,998	1,344	74,058	8,195	406,211

Source: SGRP, 1996   ** Developing countries   *** Developed countries

A Tremendous Return on Investment

- World Bank's 1972-1997 contributions of about US$600 million mobilized investments of more than US$4 billion from other sources.
- In 42 developing countries surveyed, every US$1 increase in agricultural production attributed to CGIAR generated US$2.32 of growth in the overall economy.
- The additional output of wheat produced in developing countries from CGIAR varieties is estimated at US$1.8 billion annually.
- The three-fold increase in maize production in West and Central Africa (1981-1996), due to new varieties developed by the CGIAR, can feed 40 million people each year, with a value of US$1.2 billion.
- The successful control of the cassava mealy bug, a CGIAR-led effort, added more than US$400 million annually to output in sub-Saharan Africa.

Wheat – A Major CGIAR Success

Because CGIAR researchers were able to modify wheat so that it can be grown in hot climates, not just temperate and subtropical zones, wheat will soon become the number one grain in developing as well as in developed nations.

CGIAR research is responsible for nearly three-quarters of all spring-bread wheats grown in developing countries.

Thanks to CGIAR research, average wheat yields have doubled since the early 1970s.

Rice – CGIAR's World-wide Impact

Beginning with the introduction of the first high-yielding, semi-dwarf varieties in the 1970s, research done at CGIAR's IRRI in the Philippines has had a global impact on rice production.

- 75 percent of the rice produced in Asia is of the high-yield variety developed by CGIAR researchers.
- CGIAR holds in trust the world's largest and most comprehensive collection of rice genetic resources – more than 90,000 samples.
- CGIAR researchers are working on a new breed of "super rice" that may increase yields by 25 percent and use less water.

CGIAR researchers in Africa at WARDA, building upon the rice improvement work already done by IITA, have developed totally new, low-management plant types that combine the best of the Asian and the African rice strains. They are also working with African farmers to improve methods of cultivation and rice selection.

Other Crops — A Pattern of Success

CGIAR research has also boosted the productivity of other crops that are staples for the developing world.

Cassava: Nearly 60 new varieties have been introduced, some developed to resist pests and disease.

Beans: More than 180 varieties, developed with CGIAR germplasm, have been released in the past 25 years.

Forage Grasses: Improved varieties led to an impressive increase in animal productivity, which has improved supplies of milk and meat in the developing world.

Potatoes: CGIAR scientists successfully developed

potatoes resistant to all known forms of late blight disease. Lead researcher Rebecca Nelson received a MacArthur Foundation "genius" award, for her work on fighting potato blight.

Global Impact — A CGIAR Snapshot

- More than 300 CGIAR-developed varieties of wheat and rice in use by the world's farmers.
- More than 200 new CGIAR-developed varieties of maize being grown in 41 countries.
- CGIAR work has produced improved varieties of legumes, roots and tubers, pasture crops, and other cereals.
- CGIAR's research has improved farming techniques, strategies for managing livestock disease, assessing fish stocks, protecting genetic resources, and managing natural resources effectively.
- The Green Revolution doubled productivity of such staples as wheat and rice.
- In India, wheat production on existing acreage nearly tripled, providing self-sufficiency.
- Some 50,000 researchers and scientists worked and trained at CGIAR centers.
- More than 600,000 accessions of germplasm held in CGIAR genebanks.

Seeds of Hope: CGIAR's Germplasm Storage Network Makes A Difference

In 1994, when civil war and ethnic genocide in Rwanda destroyed the country's farms and granaries – and the supply of seeds and planting materials needed to feed the country the following year – the CGIAR established Seeds of Hope. Drawing on germplasm resources from CGIAR centers around the world, this program collected and replicated planting materials that matched the varieties of beans, maize, sorghum, sweet potato, cassava, and plantain grown in Rwanda. CGIAR then provided the replacements to Rwandan farmers in time for the next planting season. Similar rescue efforts replaced the seed base lost during years of conflict in Vietnam and Cambodia, and more recently Angola and Liberia. In the wake of Hurricane Mitch, Seeds of Hope assisted the farmers of Honduras and Nicaragua where more than 60 percent of the combined agricultural area was severely damaged.

Dryland Research –

Help for "Forgotten" Areas

Centers have focused on technologies and policies to boost production and alleviate hunger in the dryland areas of the world where nearly 1.6 billion people live.

Results of CGIAR Dryland Research

- Developed new varieties of sturdy, drought-resistant, higher-yield pearl millet – a staple.
- Improved livestock production and the combined crop-livestock systems.

Resource Management –

Helping Farmers, Preserving Resources

Tropical forests, water resources, and fishing have been the focus of CGIAR researchers for some time. Several years ago, the CGIAR issued a report that called for a comprehensive effort, based on science, to attack the real causes of resource degradation and help farmers use sustainable agricultural practices to protect existing farmland. CGIAR researchers have made significant contributions to reaching these goals.

Results of CGIAR Resource Management Research

- Showed the potential of bamboo product manufacturing in China, marketable charcoal in Indonesia, and medically valuable plants in Peru and other Latin American countries.
- Worked with the Utah Climate Center to produce a new global database called the *World Water and Climate Atlas for Agriculture*.
- Contributed to a rise in worldwide fish production for the first time in five years.
- Pioneered the farming of giant clams and tropical sea cucumbers.

- Demonstrated the potential food-producing value of ponds to farmers in developing countries.
- Advanced stock improvement and control of animal diseases common to the tropics.
- Developed genetic markers to improve resistance to disease and parasites, and to improve livestock production.

Helping China's Sweet Potato Farmers

In 1987, CGIAR's CIP conducted a small training course on virus detection techniques for its own scientists. The following year, CIP scientists offered a workshop at China's Xuzhou Sweet Potato Research Center to teach the techniques for developing symptom-free tissue culture plantlets. Thirty of China's top sweet potato researchers attended.

The newly trained Chinese scientists, supported by scientists from the sweet potato center, worked for five years to adapt the techniques. In 1994 they began planting virus-free sweet potato. By 1998 more than half-a-million hectares were planted with the virus-free plants. With the new plants, yields improved as much as 160 percent and farmers had not been required to change their practices.

The 1998 Review –
A New Mission Statement

Since it was formed, the CGIAR system has undergone three extensive reviews by panels of outside experts. Each time the panels have highlighted the important contributions made by the system's network of international research centers in collaboration with their international and national partners. Each time, the panels have recommended adjustments to help CGIAR more ably meet changing global food needs and confront environmental realities.

Responding to these reviews, the CGIAR has shifted focus over the years to encompass the broader range of issues upon which food security depends. The most recent outside review was completed in the fall of 1998. The nine-member international review panel, under the chairmanship of Maurice Strong, former Secretary General of both the Stockholm Conference and the United Nations Conference on Environment and Development, spent more than 18 months assessing the 16 CGIAR research centers, consulting with partners, donors and national agricultural research systems, and with farmers and consumers in the developing world who are the direct beneficiaries of the system's research efforts.

The panel determined that the specific crop approach taken by many CGIAR centers would not be sufficient to meet the food needs of a growing population and, at the same time, protect and preserve the world's natural resource base. Instead, the panel proposed a far more holistic approach to the interlinked problems, with sustainable agriculture providing the common thread.

One of the major elements the panel identified was the need to view the food production system in the larger context of natural resource management.

The review panel determined that sustainable agriculture is, in fact, the means to eliminating poverty while protecting biodiversity, land and water, and effective natural resource management is the way of achieving it.

The third review promised to be a turning point for the CGIAR system. One of the panel's key recommendations was a new mission statement for the CGIAR: "to contribute to food security and poverty eradication in developing countries through research, partnership, capacity building, and policy support promoting sustainable agricultural development based on the environmentally sound management of natural resources." The new statement was formally adopted at the annual CGIAR meeting in the fall of 1998.

The Future

Despite the impressive achievements from CGIAR science and the continuing work of the international centers, much work remains to be done. A recent projection from IFPRI showed that, by 2020, the global demand for rice, wheat, and maize – the three crops that are the system's greatest successes – will most likely increase by about 40 percent. Almost all of that increase in demand will come from the developing world.

Most analysts believe future increases in food supplies will come mainly from improved production, since the natural resource base will not support either significant expansion of farmlands or more extensive irrigation. That means, in effect, that production on existing lands must nearly double over the next few decades at a time when water tables are falling and erosion is expanding, highlighting the need to re-hone the tools of the Green Revolution.

OTHER INDICATORS REINFORCE THAT CONCERN.
With the expanding emphasis on natural resource
management, CGIAR's future scientific achievements
will need to be evaluated in a different context than
those of the past. Obviously, techniques that improve
production, increase resistance to pests and disease, and
improve the ability of plants to grow under less than
ideal conditions will be one important key to meeting
future food needs. CGIAR will also have to pay
scrupulous attention to preservation of the natural
resource base and genetic heritage that will ensure
future sustainability.

3. PUTTING SCIENCE TO WORK FOR CHANGE

THOUGH THE CHALLENGE OF FEEDING a growing population in the 21st century appears vastly complex, three striking advances could make the task feasible: the gene revolution, natural resource management, and the information revolution. Together, these offer the potential to radically reshape the world's agricultural and food systems. The CGIAR is uniquely positioned to use these powerful new scientific breakthroughs. It has been a leader in the development of biotechnology and the techniques of natural resource management, as well as a pioneer in using information technology to meet the challenge of fighting poverty and feeding the world.

Gene revolution encompasses the new understanding of how genes work as well as the techniques and tools of biotechnology that make it possible to manipulate genetic material as never before. The information revolution includes the computing and telecommunications advances that make it possible to quickly and inexpensively reach the most remote corners of the world with complete and up-to-date information.

Used appropriately, these breakthroughs could lead to the improved productivity and more diversified crops required for future needs. It is possible, however, that these advances will not be equally available to developed and developing nations. One of the main priorities of the CGIAR, then, is to ensure that the food security of the poor is enhanced, and not impaired, by the science and information revolutions.

These advances will complement and enhance existing approaches, not replace them. Biotechnology can change the makeup of plants in ways conventional breeders can only dream about, allowing plants to, for example, grow in saline soils or remain untouched by the pesticides used to kill weeds. Biotechnology cannot, however, make depleted lands more fertile or assure water to irrigate crops. Those will require

smarter and different farming and natural resource management techniques, which CGIAR has pioneered, some that rely on new technology and some that incorporate the traditional methods farmers have used for generations.

The science of agriculture is in the throes of massive change. Classical agronomy is joining forces with the science of ecology. Research and experimentation is no longer dictated and directed from a distance but now includes local farmers as active partners. Some research no longer targets a single crop or problem, but focuses on the integrated management of natural resources. This new, more holistic approach strives for sustainability, and ecological, social, and economic viability, not simply new crops and higher yields.

Modern telecommunications technologies can play a key role in identifying priorities for research needs, in scientific and technical training, and in the support of local research and development efforts. The ability to make knowledge and information about new advances and useful technologies widely and quickly available through the Internet, cellular telephones, advanced computing systems and other means will propel change. But communication is a two-way process. This knowledge will only be useful if scientists communicate with farmers so their efforts respond to real-world needs, and if their knowledge reaches farmers in a form they can both understand and use.

Biotechnology

WHAT THE NEW SCIENCE CAN DO. Biotechnology and the genetic revolution have made striking advances in a very short time. The first transgenic plants were introduced experimentally in 1982. Since then, different combinations and varieties have been tested in all parts of the world. The first of these crops became commercially available in 1996, when somewhere between two and three million hectares were planted with genetically altered strains. By the following year, approximately 15 million hectares had been planted.

For thousands of years, farmers have selectively bred crops and animals to improve output, the same strategy used successfully by scientists at CGIAR's research centers and elsewhere. But selective breeding has aimed mainly at enhancing production and increasing the ability of plants and animals to resist disease and other stresses.

Biotechnology has added new dimensions. For example, the nutritional content of a crop has not always been a guiding concern for traditional breeders, but plants now can be engineered to improve their nutritional content and counter specific deficiencies. In many developing countries where grains are staple foods, meat and vegetables are not important parts of the diet, and micronutrient deficiencies are common, genetically engineered foods can lead to better health.

The insertion of a gene that produces beta carotene, a precursor to vitamin A, into the canola plant has provided one biotechnology success story.

Vitamin A deficiencies, common in the developing world, can cause irreversible eye damage. Since many farm families grow canola to produce their own cooking oil, this genetically altered plant offers a potential solution to a major health problem. The beta carotene content of one tablespoon of oil pressed from the seeds of these genetically improved plants is sufficient to prevent both vitamin A deficiency and subsequent eye damage. Researchers are also working to add genes to rice to boost the iron content and help prevent iron deficiency anemia.

Other gains are being made with genes that confer resistance to insects or diseases, or that counter less-than-ideal growing conditions. In India, for instance, scientists have added two genes to rice to help plants survive when submerged for long periods, a problem common in Asia. Researchers in Mexico, funded by the Rockefeller Foundation, have added two genes to rice and maize that appear to increase tolerance to aluminum to counter the soil toxicity problem common in many tropical areas.

Few question the potential contributions of biotechnology. Nigerian President Olusegun Obasanjo, while urging fellow African leaders to "invest in education, science and technology to improve efficiency and social discipline," notes that unlike the development of the 20th century, which was based on fossil fuels, the 21st century would be "biological and biodiverse, with biotechnology being the kingpin of the process."

CONTROVERSY. Despite this promise, many concerns have been raised about the potential safety of the new technology and about the ethics of moving genetic material between species. In 1998, Great Britain's Royal Society issued a statement supporting the use of genetically modified plants for food use, but with an important caveat. The technology, the

Biotechnology's Potential Contributions to Food Production

- More food produced from existing land and water resources, reducing pressures to expand farming into the wilderness, rainforests and marginal lands that are the world's richest remaining repositories of biodiversity.
- Reduced post-harvest food losses, boosting the "realized nutritional yield" of food per acre.
- Reduced need for costly and resource intensive inputs such as fuel, fertilizers, and pesticides.
- Greater use of more sustainable farming practices such as conservation tillage, precision agriculture and integrated crop-livestock-tree management.

statement proclaimed, "has the potential to offer real benefits in agricultural practice, food quality, nutrition, and health." But at the same time, the statement continued, the 'legitimate concerns' about biotechnology must be addressed, to achieve consumer confidence based on sound scientific evidence and systems of checks and balances, if the technology is to contribute to the challenge of feeding the world's growing population."

Alert to those same concerns, the World Bank named a Panel on Transgenic Crops in 1997 to examine, among other issues, the potential role of biotechnology in the challenge of providing an adequate food supply based on sustainable agricultural practices. The greatest contributions, the panel decided, would come from research targeted specifically to meet the food needs in the developing world, and from "spillover" innovations initially aimed at industrial countries which also turn out to be useful for food production in the developing world.

The analysis done by the panel, chaired by the late Nobel Laureate Henry Kendall, did not conclude that biotechnology alone would meet future food needs. It did show that efforts to improve rice through biotechnology should begin to be realized within a few years and that biotechnological improvements to tropical maize, cassava, and banana production also could be expected. In the long term, the panel concluded that efforts to incorporate pest and disease resistance into developing country crops are apt to be more successful than those aimed solely at improving yields.

The panel identified some problems and costs likely to be associated with wide application of biotechnology. For example, genes that protect plants from disease or insects can move from a domestic plant into a wild relative, making the wild plant more difficult to control, potentially disturbing the ecological balance of an area. Panel members also pointed out that biotechnologies can have both benefits and drawbacks for developing countries. For instance, tissue culture can be used to produce disease-free plants and to increase productivity in the developing world, but it can also be used to shift a traditional developing world product to the developed world. As a case in point, they noted that tissue culture work has made it possible to produce vanilla, a product of the tropics, in laboratories anywhere in the world.

Like so many new technologies before it, biotechnology has been beset by apprehension and misunderstanding. This has provoked considerable opposition to both research and application, "not all of it carefully thought out," the panel noted. But this opposition cannot be quickly dismissed, the panel cautioned – much as the Royal Society had – for "public perceptions that genetically engineered crops and animals products pose specific dangers must be

Like so many new technologies before it, biotechnology has been beset by apprehension and misunderstanding.

The questions of biosafety and the ethics of manipulating genetic material, need to be resolved before the potential of biotechnology can be realized.

carefully considered and addressed if such products are to reach widespread use."

The swings of public opinion, which create the political environment that either supports or opposes a technology, seem fundamentally different in North America, Western Europe, and Japan. In Europe, for instance, anti-biotechnology sentiment is slowing the introduction of genetically engineered crops as public opinion focuses on perceived risks rather than the benefits. Opponents have used scare tactics warning of impending epidemics of cancer, birth defects and sterility, and uncontrollable "monster" plants if genetically altered materials are used. To date, however, no incident in genetic engineering or biotechnology suggests that the risks are any different, any greater, or structurally different from those associated with classic Mendelian research.

Biotechnology also has run into fierce opposition from proponents of organic farming who seek a return to "natural" farming, using no chemical fertilizer or sprays. The organic farming movement has gained momentum, especially in the United States and Western Europe, where concerned consumers want produce and livestock untainted by manufactured fertilizers, pesticides, antibiotics, or growth stimulants.

One further problem area involves what is known as the "terminator" gene. In theory, terminator technology would force a plant to produce a toxin that is fatal to its own seeds. That would eliminate any possibility that farmers might save seed from a genetically altered variety from one year for the next year's crop. The debate over terminator genes has been highly emotional. In India, irate farmers burned several experimental fields with crops rumored to contain the terminator gene. But in fact, terminator technology is highly complex, involving several procedures. A major USA crop biotechnology company, having imposed a five-year moratorium on terminator technology, has announced it will not commercialize it.

The questions of biosafety and the ethics of manipulating genetic material, need to be resolved before the potential of biotechnology and genetic engineering can be realized. The CGIAR review panel recommended a new unit to deal with combined questions of biosafety, bioethics, and biosurveillance so that biotechnology can be safely applied in ways that are "pro-poor and pro-environment." The panel acknowledged the "many uncertainties and potential risks" associated with biotechnology, including "adverse social and economic impacts; risks to human and animal well-being; risks to the environment; adverse food safety impacts; and risks to ethical beliefs and value systems, challenging deeply held ideas on humankind and its relationship with nature." Those risks, the panel stated, must be balanced against the potential for aiding the goal of food security for the poor.

The risk-benefit equations may be quite different in developed and developing countries, for those faced with malnutrition or starvation may not be worried about the incompletely-defined theoretical health risks cited by American and European pressure groups. The poor need the tools and information to decide about the use of biotechnology based upon their own needs and priorities.

The Problems of the Developing World's Poor. Because many breakthroughs in biotechnology are the product of proprietary science, the CGIAR and other development organizations have been concerned that significant advances will not be available to address the problems of the poor. Even though scientific research is a long-term process with no guarantees of success, investors would not risk such sizeable sums unless they knew they could protect the intellectual property rights to any new discoveries so they can recoup their investments and make a profit.

That protection is necessary for this kind of research to thrive, but it raises the very real issue of whether the larger public interest, the so-called public goods, will be served by proprietary science. Past experience provides no sure answers. Computers were developed by the private sector, but competition and second- and third-generation systems have made computer technology available to the developing world. Experience with the pharmaceutical industry, also driven by private sector investment, has been different. For example, these companies have done little to develop a vaccine for malaria, which is the largest disease in the world today, but one found mainly in the developing world.

Biotechnology at CGIAR. Even as such concerns are being discussed within the CGIAR, support for greater involvement in biotechnology research has grown. In 1998, the overall CGIAR investment in biotechnology reached US$30 million, accounting for about 7 percent of the total budget. The investment in some centers is proportionately greater. CIP in Peru, for instance, has devoted 15 percent of its budget to biotechnology, while ILRI devotes 20 percent of its total budget to support biotechnology research on vaccines, diagnostics, and the genetics of disease resistance. Because it draws heavily on biotechnology research for human health problems, work at ILRI, in turn, has contributed to these efforts. For instance, ILRI researchers produced the first vaccine against a haemo-protozoan parasite, *Theileria parvia*. Their work has provided useful information for the development of vaccines for malaria and sleeping sickness.

CGIAR's Target – The Needs of the Poor.
The work at the CGIAR centers focuses on biotechnology appropriate to the needs of the poor. CGIAR researchers have developed pearl millet resistant to downy mildew and are working to develop insect resistant cowpeas. But center researchers are also using molecular marker techniques to shorten the time and reduce the costs needed to develop new crop varieties,

---

The Apomixis Gene – New Help for Developing World Farmers

In cooperation with French scientists, CGIAR-supported researchers at CIMMYT are working on the apomixis gene, which, when perfected, would give plants the capacity to reproduce asexually. Hybrid plants are vigorous, but their seeds are notably lower yielding than the original plant so that farmers can't replant seeds profitably. If farmers had apomictic versions of maize and other hybrids, they could then replant seeds from their own harvests instead of purchasing fresh seed each year.

The need to buy seed each year perpetuates a vicious cycle that has prevented many small farmers from using improved hybrid varieties and benefiting from the increased yields – which might boost their incomes enough to justify the expense of the seeds. "Apomixis fixes diversity in the wild and could be used in agriculture to retain hybrid vigor and allow poor farmers to save seed of their best plants for use the following cycle," explains CIMMYT Director General Timothy Reeves. "Hybrid seed production in apomictic maize would be simpler and less expensive. Selection of superior, locally-adapted varieties would be enhanced."

---

both in the laboratory and in the field, and identifying new tools to diagnose plant diseases.

Many farmers have discovered that using genetically altered strains changes the way they grow their crops. For example, soybean farmers who have tried the new varieties report the combination of genetically engineered crops and conservation tillage makes their work easier, cheaper, and more productive. Tractor use is reduced markedly, which saves both time and energy. Allowing crop residues to return to the soil reduces erosion and increases fertility, and also increases the ability of soil to retain water. Because of such benefits, the use of pesticide-resistant soybeans has grown from 1.5 million to more than 20 million hectares in four years.

Some developing world farmers are already sharing in the benefits of genetic engineering and appear to have profited. In China, a group of 650 small cotton growers tried seed for bollworm resistant cotton. They paid a premium for the seed, but in return they got a 90 percent germination rate, not the usual 40 percent. They saved money because they needed no spray and did not have to pay for workers to pick off the worms by hand. Three years later, 1.3 million small farmers had switched to the new seed and that number is growing.

Natural Resource Management – CGIAR's Approach

Any gains made through combinations of biotechnology and traditional breeding, though, have to be integrated with sound strategies for managing natural resources. That is why the CGIAR's crop improvement approach emphasizes the vital role that natural resource management must play in achieving the goals of sustainable agriculture.

Most natural resources are renewable, but some are not. Each can be studied independently – soils, water, rivers, coastal zones, forests, biodiversity – to understand flows and cycles. But together, natural resources form a system which has to be considered as a whole. The CGIAR's research, therefore, assesses and includes the combined stresses and replenishment capabilities of different ecosystems to achieve maximum productivity. This approach helps the CGIAR provide new varieties that are suited to conditions in a given ecosystem.

WATER. The challenges of natural resource management are system-wide, though the precise problems vary from region to region and ecosystem to ecosystem. Water provides a good example. Unless fresh water resources are properly managed everywhere, water shortages are likely to be the most severe constraint on food production in decades to come. IWMI scientists are studying all forms of water use, not just irrigation, and have found that about half the anticipated increase in water demand can be met by more efficient irrigation – which would also ease water-logging, salinity, and other problems caused by too much water.

Through a Systemwide Initiative on Water Management (SWIM), CGIAR scientists are working with national and regional water experts to identify techniques for improving water management in regions already confronting shortages, and to develop strategies to help other areas conserve and use water resources more effectively and avert future shortages. Similarly, ecoregional consortia of centers and their national partners are involved in programs that focus on livestock-related natural resource management.

Most CGIAR Centers Are Well Established and Are Able to Specialize in Certain Aspects of Resource Management.

- ICRAF has designed research projects to resolve the question of whether agroforestry systems or landscape mosaics – where monoculture alternates with plantation forests and conservation areas – are more successful for increasing yields and protecting resources and biodiversity. The two questions, or hypotheses, are being tested in different ecoregions. Results will be compared and the trade-offs assessed so national and regional systems can be advised about which approach is more effective.

- ICLARM has developed fisheries and coral reef resource assessment and database tools to enable both scientists and decision makers to assess the status of stocks, as well as the biological and socio-economic data relating to the use of a given aquatic resource. These tools make it possible to evaluate resources on a local, regional, or global scale.

- Several centers, including ICRISAT, IITA, and ILRI have looked at livestock-related nutrient cycling and documented the importance of livestock manures to small farmers to increase crop productivity and maintain soil fertility.

- CIAT scientists in the field are utilizing another resource management approach. They study and analyze the geographical, social, economic and political factors that shape resource management decisions to help communities develop better resource management strategies. Geographic information systems and mapping tools provide complex land use information that is relatively easy to use and to understand. Such information can be used at different levels to help local communities make resource management decisions and those at national levels to analyze resources and evaluate overall management options.

Effective tools of resource management can be adapted widely to meet local needs, with growing emphasis on precision farming and such strategies as integrated pest management and integrated nutrient supply systems which protect the environment and help preserve the resource base. CGIAR research is showing how natural resource management can provide the foundation needed to achieve the goals of agricultural sustainability in the developing world.

The Information Revolution

None of the advances from any of these efforts will be of value unless they can be used to alleviate poverty by those who need them most – the small farmers of the developing world. The recent convergence of computing technology and telecommunications means that vast amounts of knowledge can be sent anywhere quickly and cheaply, making available to the developing world the large databases, libraries, remote sensing, gene banks, and other sources that were once far too expensive or remote.

GREATER ACCESS AND SCIENTIFIC COMMUNICATION. Some scientific organizations have made special access to electronic scientific journals freely available to the developing world, a practice likely to spread because little cost is involved. This means that the world's scientific and technical literature will be available everywhere, in developing as well as industrial

countries, as soon as it is published.

The phenomenal growth of electronic networks in the past few years can be used to create a more interactive global agricultural research system. Unlike radio or television, the Internet facilitates two-way communication, making it possible to tap into and share the innovative talents and experiences of farmers and scientists in all parts of the world.

In the past, indigenous knowledge about local varieties, farming techniques, and other local technologies tested through the generations rarely made its way to scientists who could incorporate it in their work. Now, indigenous knowledge, combined with new and classical scientific knowledge is available worldwide.

Farmers in the developing world need more than access to the local region-specific information that can affect the common problems they encounter every day. They need, for instance, information about water systems, the use of fertilizer, soil conservation techniques, and new plant varieties best suited to specific growing conditions.

As the external review panel noted, the information and communications revolution "presents a tremendous new opportunity for the CGIAR to bring scientific knowledge and indigenous and local knowledge together to bear on global challenges, and to make this information available to its constituents." Furthermore, the panel concluded, "the CGIAR must be at the forefront of harnessing these technologies to pursue its mission."

That conclusion could also apply to the role the CGIAR must play to assure that the benefits of the gene revolution also reach the developing world. The CGIAR is well-positioned to speed the application of these two tools to the problems of the developing world. Only by doing so can the CGIAR hope to fulfill its mandate to ensure sustainable food security for the generations to come.

4. WHAT IT WILL TAKE

The CGIAR has the system in place and the tools at hand to meet the complex challenges of the looming global crisis in food and agriculture: food security, poverty, environmental protection, and genetic resource preservation. As the Third System Review Panel said, "There can be no long-term agenda for eradicating poverty and ending hunger without the CGIAR."

But the panel also found that CGIAR will only be able to do what is needed if it can expand its sources of funding and secure more predictable and stable support. That requires bold initiatives from the CGIAR and stepped-up commitments from governments, international organizations, national research organizations, NGOs, and the private sector.

Moreover, the panel identified additional specific steps the system should take both to meet the challenges and to use the available tools effectively. It emphasized that the CGIAR must find ways to make its work more visible, increasing public awareness throughout the world, of its past accomplishments and of the pivotal role CGIAR science can play in meeting the challenges of the 21st century.

Recommendation:
Financial Support – The Need for Greater Commitment to CGIAR

As the review panel outlined the specific recommendations, members also detailed the kind of support needed to make turn their suggestions into reality. Given the important role it has played in agricultural research, the CGIAR has not been an expensive system. Gains from investment in the system have far outpaced expenditures. One recent analysis, based on a large number of independent studies, found the overall rates of return averaged 73 percent.

Funding was not a major issue for CGIAR until donors and assistance agencies began curtailing contributions in the mid-1990s. The renewal effort of the mid-1990s boosted support to about US$350 million a year. With the new expanded research agenda, the review panel calculated the system would need resources of about US$400 million a year going into the 21st century, as well as a long-term funding strategy to meet the needs of the growing research agenda.

CGIAR Expenditures by Developing Region: 1998
Total $US337 Million

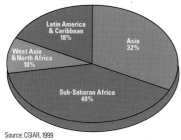

Source: CGIAR, 1999

CGIAR Member Investments: 1998

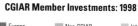

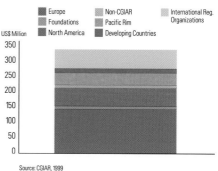

Source: CGIAR, 1999

Overall rates of return

on investment in CGIAR

averaged 73 percent.

The international

development community

must reverse the decline

in official development assistance

for agriculture and

agricultural research.

— The Third System Review Panel

At the same time, the panel found it "unrealistic" to expect that increase to come from the system's traditional funders and urged the CGIAR to diversify funding by reaching out to the private sector, the philanthropic sector, and to national development agencies. But the panel also recognized the need to maintain the strong support of traditional donors, especially the World Bank. As with many funders, the Bank provides intellectual support and legitimacy as well.

To maintain levels of past support and attract new donors, CGIAR needs to maintain its traditional strengths while demonstrating – and publicizing – both impact and cost effectiveness, the panel said. Over the years, the composition of system funding by members has been relatively stable, with nations of the developed world providing about two-thirds of the total, development institutions about one-quarter, and the rest split between developing countries and foundations.

Since most countries of the industrial world already are members of CGIAR, the prospects for new support from this important group of contributors are slim. Recently a number of developing countries have joined the system, prompting the panel to suggest the membership from this group should be doubled over the next decade, not only to provide additional income, but to give a "strong political signal" of support for the system and its agenda.

Based on its assessments, the panel recommended "that the international development community reverse the decline in ODA (official development assistance) for agriculture and agricultural research, tap other non-ODA public sources, and commit all parties (governments, international organizations, national research organizations, NGOs and the private sector) to coordinate their resources and efforts to combat the risk and threat of pervasive poverty, food insecurity and environmental degradation in developing countries. Given the challenges ahead, this is a time for greater financial commitment to the CGIAR."

Recommendation:
The Private Sector – New Cooperation, Uncompromised Objectives

One of the big question marks in the CGIAR future involves the role of the private sector. Through the years, individual centers have received funds from private sources, usually targeted to specific projects. The panel highlighted the potential for greater contributions to collaborative efforts and also for new private funds targeted to training and education programs.

> *Many in the developing world and the NGO community want collaboration with the private sector to focus on helping the world's poor.*

Many in the developing world and the NGO community, however, are reluctant to increase involvement with the private sector. They suspect collaboration is motivated by the desire to maintain access to genetic resources. These concerns prompted the review panel to recommend, as an overall policy, that CGIAR collaborate with the private sector only under conditions that do not threaten or compromise the system's overall objectives. "Financial contributions from the for-profit sector should be accepted for research activities of mutual interest, in line with the CGIAR mission statement, and directed toward the agreed research agenda," the panel recommended.

The panel also recommended that the CGIAR use a professionally run foundation to raise support from private sources. The International Fund for Agricultural Research, originally set up to create public awareness and solicit support for the system, could provide a mechanism to handle non-traditional contributions to support the research agenda, conduct public relations activities, and provide an institutional mechanism for holding intellectual property. CGIAR itself is not a legal entity and, at the moment, only the individual centers which are legal entities, have standing to hold property rights.

Recommendation:
The Role of Governments – More Emphasis on CGIAR Policy Research

Not all the pieces needed to achieve the vision of food security for the 21st century are obvious – or easily put into place. Indeed, the success of partners in the network, from the researchers at the international centers to local farmers, hinges in large part on policies put in place by both local and national governments. They also depend upon the commitment of those same partners to the overall goals of the CGIAR.

Recognizing these needs, the review panel called for greater emphasis on policy research, especially to help strengthen the capabilities of countries where a lack of public policy support contributes to the gap between the actual and potential yields in farmers' fields. This research should cover not only economic policies, but also policies that affect the environment, science, and technology.

Recommendation:
Partnerships – Applying Science to the Problems of the Poor

Science and technology have made astonishing progress over the past decades, but the distribution has been lopsided. Most of the scientists and science are found in the industrial world. This inequality has always been balanced by the fact that the talents, tools, and findings of the industrial world have been readily available to tackle problems specific to the developing world.

With the rapid growth of proprietary, investment-driven research in the industrial world, this balance has been disturbed. The achievements of CGIAR scientists and scientists in national agricultural research systems remain in the public domain, along with the genetic resources that have made so many advances possible. As those genetic resources are altered by proprietary science – by insertion of a gene that protects against disease, for instance – the new products are protected by patents and no longer freely available for use in the developing world.

> *The threat of "scientific apartheid" means it is time to design new partnerships and strengthen existing ones.*

Partnerships could provide a solution to this growing problem. Partnerships have always been an important feature of the CGIAR, but the threat of "scientific apartheid" means it is time to design new partnerships and strengthen those that already exist. These partnerships should bring together the national and international, public and private, rich and poor, and formal and informal institutions involved in agricultural research. These partnerships should be based upon the common recognition that public and private research efforts can be complimentary, and upon the understanding that the issues transcend national and regional boundaries, political differences and individual financial interests.

By working together, partners can find ways to apply new science to the problems of the poor. Without this joint effort, the free and open exchange of genetic materials characteristic of past decades will be threatened and the value of traditional knowledge and experience could be diminished, leaving a major portion out the world's population left out of the new scientific enterprise. If this happens, both the poor and the environment would be the losers.

Partnerships are essential to the CGIAR's efforts. They have a synergistic effect, expanding the impact CGIAR alone can make with limited resources. Working with national systems ensures that research tools are available and findings adopted. Partnerships with the private sector and advanced research institutes ensure that scientific advances are available to serve the public good and needs of the poor. Partnerships with NGOs and farmer groups create a two-way process that both responds to the needs of farmers and ensures the application of results. Partnerships among CGIAR centers and with other international research centers expands and extends the system's expertise.

Recommendation:
Intellectual Property – Protecting the Interests of the Poor

Until such partnerships materialize, strong, publicly-funded research is needed to deal with problems that may not concern the private sector. So far, commercial firms have shown little interest in plants, pests, and diseases common to tropical areas. Their products have been more suited to large-scale, commercial farming than to the complex systems typical of rural farmers who may mix food crops with cash crops, livestock, agroforestry, and even aquaculture on a small plot.

The situation calls for a continuing dialogue between the private and the public so that the needs of the poor are not ignored. Several large corporations, alert to these needs, have donated money, research and products to developing world research – one firm providing a key gene for sweet potato research in Kenya, another a Bt gene to IRRI – and that is a promising start.

> *Private investment in biotechnology has grown at a staggering pace.*

Intellectual property rights and issues of farmer access and ability to pay lie at the core of the tension between commercial interests and the needs of the poor. Advocates for intellectual property rights believe that inventors have a right to protect and recoup their investments, and that this financial incentive fuels further research. Those who oppose the protection worry that restrictions block open exchange of resources and the sharing of knowledge. This, they argue, shifts control of knowledge and resources from farmers and local communities to large corporations. It also tends to skew the research agenda to favor industrial agriculture, bypassing the community and diversity-based farming that, over the generations, has fostered and maintained genetic diversity.

Private investment in biotechnology has grown at a staggering pace and multi-national corporations with the resources to pursue the research and the expense of

patenting are dominating the field.

Precisely which patents in agriculture will be upheld, once the dozens of infringement lawsuits are resolved, is unclear. The parameters should start to emerge, but already different countries have different rules. In the United States, some corporations have protected their patents guaranteeing no seed is saved for future use. In Europe, in response to the mounting public sentiment against biotechnology, certain techniques and advances are not allowed. Many countries in the developing world do not permit patenting of plants and animals. To further confuse the situation, dozens of companies now hold two or more patents for different uses of the same gene.

The main worry for the developing world is that the added costs of patent protection could block access to many new technologies, both processes and genetically altered plants. It could also delay the availability of new probes and promoter sequences, indeed almost every new tool of biotechnology. In theory, the use of processes and transformed varieties can be blocked for the life of a patent – as long as 20 years.

The review panel urged the CGIAR to create a legal entity to hold and manage international patent rights on behalf of the centers. Any move to patent the scientific achievements of the CGIAR centers would not be competitive, for the use of international patent rights would be allowed under free license, assuring the public nature of the CGIAR patents.

The panel called its recommendation defensive, reflecting the need to maintain a strategic place for public research directed towards the reduction of poverty. At the same time, the panel urged the CGIAR to participate in the development of "rules of engagement" involving both the public and private sector, based on the premise that access to the means of food production is as much a human right as access to food.

Recommendation:
Genetic Resources – An Integrated Gene Management Initiative

The vast store of germplasm maintained by the CGIAR centers, coupled with the availability of different locations and conditions for testing new varieties, underpin the system's crop breeding achievements.

> *Many of the world's greatest advances in food production have included material from CGIAR collections.*
>
> — Geoffrey Hawtin, Director General of IPGRI

To preserve and enhance this strength, the review panel urged the CGIAR to undertake an integrated gene management initiative. The initiative would extend beyond the conservation of plant genetic materials to encompass policies for biosafety and gene deployment techniques that avoid any risks associated with the release of genetically altered materials. The proposed initiative would extend conservation to include both the genetic resources held in banks and those maintained *in situ* on farms, and it would enlarge the CGIAR "marketbasket" of crops to include minor crops that are of little value in the developed world, but essential for the nutritional security of many households in the developing world.

To assure the long-term success of the preservation effort, the review panel also called for a special fund that would serve to insulate the germplasm collection from the inevitable funding uncertainties that will confront the CGIAR in the future. This fund is crucial if genetic resources are to be available for current and future use.

> *The CGIAR should participate in the development of "rules of engagement" involving both the public and private sector.*

Recommendation:
Natural Resources – More Careful Management

The protection of plant genetic resources that are the food heritage upon which the future depends extends beyond resources stored in gene banks. It includes resources protected by farmers in local communities and those that still remain in nature, an unknown number of them undiscovered and unidentified. The review panel warned that failure to manage natural resources would lead to large, though unquantifiable, losses of genetic resources. The same failure to manage resources effectively has also meant that the yields from farmers in many areas fall far below their potential.

The review panel called for "much more careful NRM work than ever before" and proposed ways to make CGIAR's commitment to sound management of natural resources more effective. Because resource management challenges are defined by ecology and region, not by national and local boundaries, the panel suggested an international and integrated approach beginning with a systematic analysis to identify existing problems, possible approaches, and solutions.

Both problem identification and solutions would begin and end at the local level, with CGIAR scientists and their partners working directly with local farmers

> *Failure to manage natural resources would lead to large, though unquantifiable, losses of genetic resources.*

and communities. The panel recommended that the centers refocus both research efforts and partnerships to give greater emphasis to natural resource management as an essential tool in their efforts to develop a sustainable and more productive system of agriculture for the developing world.

Recommendation:
Information Exchange – A Global Food Security Knowledge System

The usefulness of gains made through biotechnology, sustainable agriculture, and natural resource management will depend upon how well knowledge about these gains is communicated to the farmers who will put it into use. To provide information to those who need it, the panel emphasized the need for an effective Global Knowledge System for Food Security that would include both cutting-edge science and technology, and traditional knowledge.

In the course of their analysis of CGIAR, members of the panel visited many of the multilateral, bilateral, NGO, and national programs for improving sustainable agricultural productivity in different parts of the world. "Everywhere we went, we found a surprising lack of knowledge and information exchange among these efforts," the panel noted. As a result, opportunities for cooperation were missed.

The proposed global system would take full advantage of the revolution in communications technology. It would eventually develop specific Web

sites to assist poor farmers with specific problems, offer virtual on-line courses anywhere in the world, and make the latest scientific data instantly available.

Recommendation:
National Agricultural Systems – Building Stronger Partnerships

The responsibility for building strong national agricultural systems, as part of overall development strategies also lies with individual governments. The CGIAR can, however, encourage governments to provide financial and institutional support for these systems. "Farmers and the national systems must remain the cornerstone of all future research efforts of the CGIAR," observes Cyrus G. Ndiritu, director of the Kenya Agricultural Research Institute. Further, Ndiritu believes, "reinforcement of the position of the national systems through equal partnerships with the CGIAR and based on mutual trust, respect, and confidence as well as capacity building and communication will be the key to the whole development process."

> *A new global knowledge system for food security would help the developing world take full advantage of the revolution in communications technology.*

In the past CGIAR has strengthened national systems by providing training, information, and research materials. CGIAR's centers have also helped to link the national systems with agricultural research institutes in other nations and to link national systems in cooperative efforts. For example, IRRI has worked with national systems in China, Vietnam, Cambodia,

Egypt, Madagascar, Thailand, Bangladesh, Indonesia, and the Philippines to help these countries develop their own rice research programs. The linkages have also benefited CGIAR researchers who have learned about local growing constraints, as well as the economic, social and ecological environments in which farmers work, providing useful "bottom-up" guidance for future research.

> Regional research organizations created by countries with common agricultural problems should receive increased support.

The financial constraints of recent years have changed the CGIAR's role. Today the primary role is to facilitate collaborative research, providing access to research materials, and management support. Training, once an essential component, has become limited to the "training of trainers." The review panel suggested that CGIAR's international centers continue to provide direct training for scientists and technicians from the weaker national research centers, and also provide training in the techniques of gene-based research and in the bioethics and biosafety considerations surrounding the new biotechnology.

The review panel also urged increased support for regional research organizations that have been created by countries with common agricultural problems, and it urged the CGIAR to "become more adept at generating bottom-up, demand-driven projects in which the prime potential users of new knowledge including regional research organizations, farmers groups, and NARS have

been engaged fully in planning phases, making it clear they have some ownership in the outcome."

CGIAR must also help national systems expand their research efforts to include natural resource management and the protection of biodiversity, as well as such food-security related issues as poverty reduction and the development of information systems. The panel suggested CGIAR should also involve the non-profit and for-profit research systems, universities, NGOs, and other international agricultural research centers in the effort to strengthen the national systems.

Recommendation:
Food Security in Africa – A Capacity Building Initiative

Food security in Africa remains a major challenge. Gains in sub-Saharan Africa have consistently lagged behind potential, leading the review panel to call for a special Inter-Center African Capacity Building Initiative for Sustainable Food Security to stimulate agricultural progress in the region. Unless constraints to higher yields are overcome, one-third of the population of sub-Saharan Africa will not have sufficient food by 2010, according to CGIAR estimates.

Population and Food Production Indices: sub-Saharan Africa

Source: IFPRI, 1997

Recommendation:
The CGIAR System – Strategies for Better Focus, Integration and Direction

The panel recommended that the international centers that comprise the CGIAR network cooperate more to take better advantage of complimentary strengths and expertise. This will mean strengthening links between the centers themselves, not just each center's links to its national system and other CGIAR partners. Because the system is relatively small, the panel suggested the centers avoid duplication of efforts. This could mean some consolidation of existing programs. The panel emphasized that research needs to be coordinated and well-focused to meet the specific needs of farmers in the developing world and that priorities must be established through close collaboration with the ultimate beneficiaries, the farmers, to ensure that research responds to their real needs.

For the CGIAR itself, the panel suggested a number of measures to both streamline operations and enhance effectiveness. By modernizing its operations, the panel said CGIAR would be able to respond quickly to changing demands and to identify research priorities as they develop, then target resources to those priorities. But much of the success of the system in meeting its goals, the panel concluded, will rely on its ability to demonstrate and then publicize the important role international agricultural research plays in the coming effort to feed a growing population, reduce poverty, and protect natural resources.

CONCLUSION

ONLY THE CGIAR can deal with agricultural research problems that cut across boundaries and affect entire regions and ecosystems. But the CGIAR cannot do it alone.

The world enters the 21st century on the brink of a new food crisis that is as dangerous, but far more complicated than the threats it faced in the 1960s. The United Nations projects that an additional two billion people, or more, may need to be fed by 2020. Even today, in times of apparent abundance, the FAO reports that some 840 million people do not have sufficient food. Millions more suffer "hidden hunger" in the form of protein and micronutrient deficiencies.

Increased affluence and a growing middle class in many developing countries, as well as in developed nations, mean that food demands of the next century will not be satisfied simply by increasing staple food production. The dietary changes – especially the increased consumption of animal protein – and the growth in nutritional intake that accompany prosperity will make the needed increase much greater than the raw numbers of projected population growth would indicate.

At the same time, the need for food often threatens natural resources as people strive to get the most out of land already in production or push into virgin territory for new agricultural land. The damage is increasingly evident: water shortages; arable lands lost to erosion, salinity, desertification and urban spread; disappearing forests; threats to biodiversity; and overfished waters. Superimposed on these are the dangers of global warming and changes in climate and growing conditions throughout the world.

Without question these threats are global and demand the attention and commitment of all peoples

and countries. Because its international status protects it from political pressures and regional or national influences, the CGIAR is in a unique position to deal with agricultural research problems that cut across boundaries and affect entire regions and ecosystems. It has flexibility and autonomy and, through its network of centers, works cooperatively with both developing and industrial countries. It has a 30-year record of accomplishment.

The CGIAR has the will – indeed, the mandate – to pursue the areas of special concern to the developing world, where the private sector is less likely to invest. While many international groups and organizations are involved in agricultural research, the CGIAR directs its attention to such developing world issues as household security crops like cassava, environmentally sound technologies that can be used by farmers on marginal lands, the diseases of tropical dairy cows, and training scientists from the developing world to work on the problems specific to the developing world. Through its unmatched networks and partnerships, the CGIAR can draw upon the expertise and talents of scientists throughout the world.

This unique commitment is needed to ensure that the benefits of new technologies are available to the poor. Many small farmers missed out on the benefits of the Green Revolution because of their limited capacity to take risks and their marginal resource base. Without the support of a group such as the CGIAR, they could be bypassed by the Gene Revolution and its promise.

As the Third System Review Panel said, "There can be no long-term agenda for eradicating poverty and ending hunger without the CGIAR."

But the CGIAR cannot do the job alone. In fact, the CGIAR cannot help meet intricate global challenges unless it can expand its sources of funding and secure more predictable and stable support. At a time when the threat is increasing, support for agriculture research has been declining. Finding the resources needed to help feed the world requires bold initiatives from the CGIAR and stepped-up commitments from governments, international organizations, national research organizations, NGOs, and the private sector.

We have to recognize that we have a serious problem that must be addressed. We have a false sense of security because there's an abundance of grain being produced. We tend not to see the real problems of hunger in developing countries.

– Rockefeller Foundation President Gordon Conway

PART II

The CGIAR System:
A Closer Look

1. HISTORY, STRUCTURE, AND FUNCTION

Understanding how the CGIAR functions helps to explain why members of the Third System Review Panel placed such faith in the ability of this international network to deal with the multiple challenges confronting the world.

HELPING THE POOR. The CGIAR system is targeted on helping the poor through research into a broad range of food commodities and natural resource management under different climatic and geographical conditions. The CGIAR has been a major catalyst for development and research since its inception. It plays an important non-political and non-governmental global leadership role that is based on the system's acknowledged scientific contributions and extensive partnerships with national systems and with agricultural and resource management experts throughout the world.

ORIGINS. Though founded in 1971, the CGIAR's roots can be traced to the early 1940s when then-Vice President of the United States Henry Wallace, a former Iowa corn grower and Secretary of Agriculture, proposed that researchers find ways to raise the yields of major food crops in less-developed countries. The effort, he believed, would assure an ample food supply and improve living conditions in less-developed nations. Wallace eventually persuaded the Rockefeller Foundation, which at the time supported health projects in Mexico, to add agricultural research to its agenda.

Once that commitment was made, the foundation sent a small team of research scientists to Mexico in 1945. Among the early members was Norman Borlaug, who took charge of the wheat breeding program, and later was awarded the 1970 Nobel Peace Prize for his pioneering research. Scientists at what became the International Maize and Wheat Improvement Center in Mexico City – known by its Spanish acronym CIMMYT – developed a number of higher yielding varieties. After years of experimental work and field testing, the new varieties became available for commercial use. Packaged with fertilizer and directions for better crop management, the new seeds were planted widely and production increased markedly.

THE FOUNDATIONS OF THE "GREEN REVOLUTION." The success with wheat produced the first half of what has come to be known as "the Green Revolution." Higher yielding varieties of rice developed by scientists at the International Rice Research Institute (IRRI) completed the revolution. The Institute was established in 1960 in Los Banos, the Philippines, by the Rockefeller Foundation as a cost-effective way of concentrating both the scientific talent and equipment needed to increase rice productivity in ways similar to those being achieved with wheat. The Ford Foundation joined with the Rockefeller Foundation in the new effort. Scientists at the new Institute quickly amassed a collection of 10,000 rice varieties. Within four years, after multiple crossings of tall and dwarf varieties, researchers developed new commercial strains of high-yielding rice. The new high-yield rice was quickly adopted throughout the rice-growing regions of Asia.

After these initial successes, the two foundations expanded their horizons. In 1967, they launched the International Institute of Tropical Agriculture, IITA, in

Ibadan, Nigeria, to develop sustainable production systems for the humid lowland tropics, concentrating on such crops as soybeans, maize, cassava, cowpeas, banana, plantain and yams – all dietary staples in these regions. The same year they opened the International Center of Tropical Agriculture (CIAT) in Columbia where, again, the focus was on sustainable tropical land use, this time in Latin America. CIAT concentrated on such staple crops as beans and cassava, forages and pasture, and rice for Latin America and the Caribbean nations.

By the late 1960s, with four centers already established, leaders of the two foundations decided the research effort needed much broader support to achieve their stated goals of alleviating poverty and world hunger. As a first step, they sought and received funding from development agencies in the United States and Canada. Then, in 1968, representatives of the two foundations met informally with leaders of the United Nations Development Program (UNDP), the Food and Agriculture Organization (FAO), and the World Bank at the Rockefeller Foundation's Center in Bellagio, Italy. Representatives of aid organizations in Great Britain, Sweden, Canada, and the United States, the Asian and Inter-American development banks, and Japan's Ministry of Foreign Affairs joined these leaders.

Robert McNamara, then head of the World Bank, proposed a consortium or consultative group to coordinate an international network of research centers, starting with the four existing centers. These centers would work closely with regional and national agricultural research programs and other agricultural experts throughout the world.

NEW CENTERS ADDED. The first four centers were so productive that seven more centers were soon added to the CGIAR network. During the 1980s, as new

The CGIAR is remarkable because it is perhaps the most successful partnership in the history of development.

– World Bank President James Wolfensohn

To contribute to food security

and poverty eradication

in developing countries

through research, partnership,

capacity building, and policy support

promoting sustainable

agricultural development

based on the environmentally

sound management

of natural resources.

— CGIAR's 1998 Mission Statement

challenges in food and agriculture emerged, additional centers were added.

As the system grew, the emphasis changed. To the initial goal of increasing productivity of food staples, which had produced the Green Revolution, the system's founders quickly added the eradication of poverty.

During the 1980s, the mounting concern among industrial nations about the deterioration of the environment led to another shift. Centers placed more emphasis on techniques of sustainable agriculture and on strategies to protect the world's remaining forests, improve livestock production, conserve water and prevent further contamination, manage threatened marine fisheries more efficiently, and protect biodiversity and genetic resources.

Another shift was caused by the realization that some countries could not consume all they produce, while others would never be able to produce enough to meet their needs. Countries were urged to concentrate on crops they could grow most efficiently and export surpluses, using the income to import other needed foods. That changed the overall emphasis from food self-sufficiency to food self-reliance, and that, in turn, expanded the research focus to include other staple crops.

The change was made with the expectation that financial resources would be available to support the new agenda, but the expanded vision came at a time when resources were uncertain. It was also made with the expectation of greater inter-center cooperation and interaction, which were slow to materialize.

The CGIAR again expanded the official agenda, committing the system "to contribute to sustainable improvements in the productivity of agriculture, forestry, and fisheries in developing countries in ways that enhance nutrition and well-being, especially of low-income people." The CGIAR also added research centers dedicated to forestry, fishery, and water issues.

Again, the shortage of resources proved a stumbling block. Despite adding new centers to deal with low-yield, high-risk areas, the system found it difficult to respond fully to these problem regions. Lack of a coherent system-wide research strategy coupled with intensified competition among the centers for funds compounded the financial difficulties.

The CGIAR is usually described as a "system," but it more aptly should be described as an informal federation of independent centers, or an arrangement for consultation. The system has two components, the 16 centers themselves and the consultative group composed of the co-sponsors, developed and developing country members, foundations, and representatives from key advisory groups and the secretariat.

The CGIAR has no constitution, only an extensive set of informal procedural guidelines. Each of the international centers, however, is a legal entity in the country where it is located and receives all the tax benefits and privileges accorded to international organizations. Each center also has its own board of trustees and its own international network linking its scientists with other scientists and groups involved in similar research activities. CGIAR, then, is actually a network of interlinked centers and their networks.

REVIEW AND RENEWAL. In 1994-95, when the system was working through a major financial crisis, a ministerial level meeting was held in Switzerland. Participants concluded that the system needed to become more of an "integrated South-North enterprise based on a shared vision."

Given the problems confronting the system and the fact that the CGIAR had more than doubled in size since the second review in 1981, a new system-wide

review seemed warranted. CGIAR Chairman Ismail Serageldin named Maurice Strong, who had represented Canada at some of the key meetings when the CGIAR was established in the late 1960s, to head an in-depth assessment. Panel members were to visit all 16 centers and meet with a wide sampling of the system's stakeholders, partners, and investors.

As had the two earlier system reviews, the third review promised yet another turning point for the CGIAR system. As a key recommendation, the panel proposed a new mission statement for the CGIAR "to contribute to food security and poverty eradication in developing countries through research, partnership, capacity building, and policy support promoting sustainable agricultural development based on the environmentally sound management of natural resources." The new statement was formally adopted at the annual CGIAR meeting in the fall of 1998.

The CGIAR research budget is a small fraction of the total for agricultural research. In non-inflation-adjusted dollars, funding has increased from the US$14.8 million allocated to the original four centers at the start, to US$141 million by 1980, to US$305 million by 1990, and to US$340 million in 1998. However, real investment in the CGIAR has declined since 1990. This decline has made it impossible to maintain consistent funding levels for each center, or to provide for any kind of growth at some of the centers.

Sources of funding have changed over the years. The two founding foundations, Rockefeller and Ford, now account for less than 2 percent of the total CGIAR budget, with countries of Europe providing about half, the World Bank 13.5 percent, the United States about 12 percent, and Japan 10 percent. Canada, Australia, and Colombia also have made significant investments. The remaining funds have come from several of the developing countries.

By mid-1999, the CGIAR partnership consisted of 58 industrial and developing countries, private foundations, and regional and international organizations that provide financial assistance and technical support. Numerous other public and private organizations work with the CGIAR as research partners, advisors, and donor investors.

2. THE CGIAR SYSTEM'S ACCOMPLISHMENTS

OVER THE PAST THREE DECADES, the CGIAR has established an impressive record of accomplishments in all regions of the developing world, contributing greatly to global food production.

Wheat

Nearly three-quarters of all spring-bread wheats grown in developing countries are either varieties developed at CIMMYT or varieties developed by national programs using genetic materials provided by CGIAR centers. With these new varieties, average wheat yields have doubled since the early 1970s.

Although CIMMYT researchers fear the new varieties could be approaching a "yield barrier," a large gap remains between what technology has provided and what farmers are using. Most wheat farmers in industrialized countries can put to use almost every advantage technology offers; farmers in many developing countries lag behind the technology curve.

Collaborative work by researchers from two centers, CIMMYT and ICARDA, has led to new spring wheat varieties suited to growing conditions in West Asia and North Africa. The improved varieties now account for about 90 percent of the spring-bread wheat grown in those regions and for significant production increases.

In Egypt, for instance, yields increased by as much as 33 percent between 1994 and 1998 as a result of the new varieties – leading to an increase in net farm income of up to 56 percent.

Rice

The widespread increases in wheat yields were achieved largely by identifying the varieties which would perform most reliably under different growing conditions. But increasing rice yields proved a different and even greater challenge. Strains that do well in Asia, where fields are extensively irrigated, do not do well on rainwater alone, which is how rice in much of Africa is grown. There are also differences in the way the crops are sown and cultivated.

Beginning with the introduction of the first high-yielding, semi-dwarf varieties in the 1970s, research done at IRRI in the Philippines has had a global impact on rice production. Not only does the International Rice Genebank at IRRI hold in trust the world's largest and most comprehensive collection of rice genetic resources, with more than 90,000 samples, but fully 75 percent of the rice produced in Asia is of the high-yield variety developed by IRRI researchers. "This is the mission of my life," comments IRRI researcher and World Food Prize Laureate Dr. Gurdev Singh Khush, "to continue to work towards the improvement of rice, and to be able to feed more and more people."

Scientific breakthroughs will yield large production increases in many developing countries where farmers now lag behind the technology curve.

The Challenge in Africa

Unfortunately, the successes in Asia have not been duplicated in Africa. The high-yielding Asian rice varieties that have replaced many of the traditional African varieties are unable to withstand rice growing conditions in Africa. Researchers at WARDA, building upon the rice improvement work already done by IITA, decided to combine the Asian and African strains in hopes of developing new plant types that might give improved yields under typical African growing conditions. Starting with 1,500 different African strains, they started a program to cross them with selected strains from the IRRI collection that seemed to have the needed characteristics.

Their efforts finally produced promising candidates. The researchers developed totally new, low-management plant types that combine the best of the Asian and the African rice strains.

The Asian experience is light years removed from what has been done at WARDA. A fundamental difference is that WARDA is developing technologies that are adapted to the African environment, without modifying the environment to fit the technology.

— WARDA Director General Kanayo F. Nwanze

Taking their new plants beyond research fields in a region with poor extension support services proved the next challenge. In a new partnership arrangement, WARDA researchers involved farmers in a participatory rice selection program using experimental plots. The farmers — mostly women, who are responsible for about 80 percent of Africa's rice cultivation — then picked out their own favorite varieties to plant on their own farms the following year.

The participating farmers tended to select varieties with rapid early growth and superior ability to compete with weeds. This produced higher yields than the original strains, as well as better ability to survive during dry periods and resist insects and diseases. This novel partnership arrangement, with the enthusiastic acceptance by farmers and extensive farmer-to-farmer seed exchange, helped spread the new varieties through all of WARDA's 17 West African member states.

Super Rice

While African scientists were tackling their particular problems, researchers at IRRI had come up with a new breed of "super rice" that promised to produce 25 percent more grain on the same amount of land.

Some of the world's best rice land is being lost to population growth and urban expansion, and remaining lands are not suited to rice production. The only way to boost rice production for future needs,

Given the growth of cities

and their demand for water,

it is unrealistic to think

that rice farmers

will get the same amount of

water for the paddies

that they have in the past.

— IRRI Director General Ronald Cantrell

therefore, is to increase yields from existing lands. But in the future, this will have to be done with less water.

Anticipating this limitation and its implications for rice growing, IRRI researchers have begun studying such water-saving ways to grow rice as "wet-seeding," or sowing pre-germinated seeds directly into muddied fields, and intermittent irrigation. "Wet-seeding has been demonstrated to be a promising technology for meeting the water needs of rice. There will also be need for subsequent changes in the management of nutrients and weeds if the full benefit of this technology is to be achieved," reports Ronald Cantrell, IRRI Director General.

Other Crops

CGIAR's efforts to boost productivity now includes other crops that are staples for the developing world. For example, nearly 60 new varieties of cassava, some developed to resist pests and disease, have been introduced as a result of CGIAR research. "IITA's campaign for the biological control of the cassava mealybug is a good example of safe plant health methods," explains IITA Director General Lukas Brader, who places the total benefits from controlling this pest at between US$9 and $20 billion. The annual value added to the crop in sub-Saharan Africa, alone, is an estimated US$415 million. Considered a "crop of last resort" for many poor farmers, cassava grows well in poor soil with limited rainfall.

CGIAR researchers have developed improved bean varieties, too. More than 180 varieties, developed with germplasm from CIAT collections, have been released in the past 25 years.

The new varieties account for about 40 percent of Latin America's total bean production, reports CIAT Director General Grant Scobie. "The new varieties markedly increase yields, resist disease, and tolerate drought and infertile soils. As a result, Latin American bean production has increased by nearly a third over the last decade," he says. CIAT research also identified improved forage grasses, which are now widely grown in Latin America. Those have led to an impressive increase in animal productivity, which has, in turn, improved supplies of milk and meat.

> In 1998, scientists at CIP successfully developed and cloned potatoes resistant to all known forms of late blight disease.

Potatoes, too, have been of special interest to the CGIAR. A 1997 study by CIP showed that potatoes and sweet potatoes accounted for a much greater percentage of food grown in developing countries than previously thought. The study also indicated that the annual growth rate of 4 percent showed no sign of abating. Production increases were identified even in some of the poorest regions of Asia and Africa.

In 1998, scientists at CIP reported they had successfully developed and cloned potatoes that were expected to be resistant to all known forms of late blight disease, the pathogen that precipitated the Irish potato famine in the 1840s. This was an especially important advance, because a new and highly virulent form of the disease has attacked potato crops everywhere in the world, except Australia, cutting global potato production by at least 15 percent.

The spread of this disease coincided with the increase in potato production throughout the world. "Growth in potato production has nearly doubled in

the last 20 years," according to CIP Director General Hubert Zandstra. "If growth trends continue at the current level, potatoes could provide enough food to sustain 250 million people by the year 2020 — more than a tenth of the total estimated population increase." He adds that finding a new potato with lasting resistance to late blight is important for feeding the developing world and protecting the environment.

A "Genius" Award

The CIP late blight achievements were recognized in 1998 when lead researcher Rebecca Nelson, a molecular plant pathologist, received a John D. and Catherine T. MacArthur Foundation "genius" award. The award cited both her concern with improving the production of staple food crops in underdeveloped countries and her efforts to disseminate the results of her research directly to farmers. Nelson considers her work a blending of research with the mission of the CGIAR "to do something about the ills of the world. I like to investigate and understand the biology of a situation," she adds, "but I also want to look at the human context."

Dryland Research

Two CGIAR centers — ICRISAT and ICARDA — have focused on technologies to boost production in the dryland areas of the world. Nearly 1.6 billion people, some one-quarter of the world's population, live in arid and semi-arid countries. These areas have attracted little attention from groups interested in improving agricultural technology for several reasons. Potential markets are small and, perhaps more importantly, researchers are still developing strategies to improve production on lands where water scarcity limits plant growth.

Since nearly half

of the world's poor people

live in dryland regions with

fragile soils and irregular rainfall,

any serious attempt

to eradicate poverty

must focus on these areas.

CGIAR researchers are looking for ways to improve the main crops of these regions by developing new breeds of plants and animals that will grow faster and stronger, need less water, and are genetically selected to provide high levels of nutrition. Their successes to date include a variety of barley that almost doubled grain yield in one of the Syria's driest areas, and several new varieties of pearl millet that combine higher yields with resistance to downy mildew, tolerate drought, and provide higher levels of protein.

> *Millions of people in the dry lands of Asia and Africa depend on pearl millet since it is the only cereal that will grow practically anywhere.*
> — ICRISAT Director General Shawki Barghouti

Dryland researchers also have worked to improve livestock production and the combined crop-livestock systems so typical of these dry lands, which are unsuitable for many crops but do support sheep and goats. One ICARDA project showed how sheep could be used to transfer legume seed from improved fields to nearby marginal lands. The simple expedient was to leave the sheep to graze on the improved pasture during the day, then move them at night to the degraded field. Legume seed, researchers discovered, passes undigested onto the land where it germinates and grows, gradually improving the degraded land.

Working with local and national agricultural research systems, ICARDA researchers have developed technology for making feed blocks by mixing, baking and pressing together locally available agro-industrial by-products. These include rice bran, date pulp, olive cake, poultry waste, and whey. The blocks are inexpensive, easy to transport, and provide a valuable food source for livestock in the region. An added benefit is that the blocks combat the degradation of rangelands and loss of biodiversity that comes from overgrazing.

ICARDA, located in the Near East, often thought to be the birthplace of agriculture, still works with some of the crops – barley, wheat, lentils, peas, flax, and vetch – that archaeologists believe were first domesticated in the region. Director General Adel El-Beltagy reports that through the combined efforts of ICARDA researchers, working with researchers from national programs, more than 230 new varieties of such ancient crops as barley, lentil, and pea, along with faba bean, bread and durum wheat, chickpea, and forage legumes have been released to both developing and developed countries.

IFPRI has placed high priority on research to assist national governments and international institutions in developing strategies and policies for dryland areas. Such research, undertaken in close collaboration with other centers, is helping governments in Latin America, the Middle East, and Africa identify strategies for dryland areas that also incorporate policies to help alleviate poverty in those areas.

Since nearly half of the world's poor people live in dryland regions with fragile soils and irregular rainfall, "any serious attempt to eradicate poverty must focus on these areas," explains IFPRI Director General Per Pinstrup-Anderson. "While outmigration may be a solution in a few cases, the sheer number of poor people in those regions make outmigration at best a very long-term solution. In the meantime, governments must seek ways to expand broad-based economic growth and alleviate poverty, which means appropriate agricultural technology for small farmers in those areas must be developed and put to work along with policies that help expand non-agricultural income generation."

Resource Management

The preservation of tropical forests has been a central focus for CGIAR researchers for some time. Researchers in Kenya identified a leguminous fodder tree, calliandra, which can substitute for expensive commercial meal to feed dairy cows. By using these trees, ICRAF Director General Pedro Sanchez explains, "a typical farmer can boost income by more than US$150 per cow each year." In Kenya, with its estimated 400,000 small-holder dairy farmers, the potential gains from this one tree amount to more than US$100 million a year. Similar benefits from calliandra could be realized by dairy farmers in Ethiopia, Tanzania, Uganda, and Zimbabwe, where growing conditions are similar. Using calliandra and other leguminous trees, IITA has demonstrated the effectiveness of the technique known as alley farming, which combines leguminous trees – which fix nitrogen from the air and extract other nutrients from deep soil layers – with food crops.

> *Forest products can provide people with good livelihoods without causing ecological destruction.*
> — CIFOR Director General Jeffrey Sayer

One center, CIFOR in Indonesia, has attempted to quantify the environmental benefits – both from limiting damage to the forest habitat and from additional sequestration of carbon – of reduced-impact logging. From this work, researchers have developed guidelines to help timber concessionaires and forest stewards manage tropical forests more sustainably.

CIFOR also analyzed the income potential of other forest goods. CIFOR research has shown, for instance,

Today, one-quarter of the

fish consumed comes from

aquaculture systems.

Within a decade, aquaculture

could provide nearly 40 percent

of the fish in the human diet.

the potential of bamboo product manufacturing in China, marketable charcoal in Indonesia, and medically valuable plants in Peru and other Latin American countries.

Water. Another relative newcomer to the CGIAR system, IWMI has worked with the Utah Climate Center to produce a new global database called the *World Water and Climate Atlas for Agriculture*. Available on CD-ROM and on the Internet, the atlas provides a high-tech tool for farmers, agronomists, engineers, conservationists, meteorologists, researchers, and policy makers, since it permits users to extract such critical data as precipitation and its probability, as well as maximum, minimum, and average temperatures for any square-mile region of the globe, no matter how remote.

"The Atlas will show where rice or potatoes, or any crop, could be grown where they are not now being grown," IWMI Director General David Seckler explains, "and it will show what more valuable or nutritious food crops farmers might grow on their land." He predicts that "the Atlas will assume even greater importance in coming years as water, especially for agriculture, becomes scarcer."

Fisheries. According to CGIAR projections, aquaculture is already a significant addition to the global food production system. Within a decade, CGIAR calculates, aquaculture could provide nearly 40 percent of the fish in the human diet, with the next major increase in food production coming from "domesticated" and genetically improved varieties of fish and other seafood.

Already tilapia is being raised commercially in more than 85 countries. "Tilapia farming has contributed to a rise in overall fish production for the first time in five years," reports ICLARM Director

General Meryl Williams. ICLARM research shows that genetically improved tilapia can grow 60 percent faster, with better survival rates, than other farmed strains and can yield three crops a year, not two. "These results are just a beginning to the production and quality gains that can be achieved for fish," says Williams.

ICLARM is also working on the genetic improvement of carp. In many locations, however, the choice of species to grow is limited and culture methods are still being developed for new species. For example, ICLARM has pioneered the farming of giant clams and tropical sea cucumbers in remote island countries. Overall, though, with the spread of new species and the improvements in productivity, the farming of fish, shellfish, and seaweeds is growing at better than 9 percent a year, by volume.

On a local level, ICLARM researchers are working on pilot projects in such diverse areas as the Philippines, Bangladesh, Ghana and Malawi, to demonstrate the value of ponds to farmers. With ponds, farmers are able to produce vegetables and other crops, as well as fish, generating a 50 percent net increase in income and providing crops in drought years. ICLARM's particular focus is on technologies and programs that reach the poorer farmers, especially women.

LIVESTOCK. Since livestock provides meat, milk, manure, and power for many poor farmers, ILRI has emphasized stock improvement and control of animal diseases common to the tropics. Trypanosomiasis, known as sleeping sickness in humans and nagana in cattle, threatens the health of both humans and animals in an area in Africa the size of the continental United States, and costs African farmers an estimated US$5 billion in lost animal productivity every year. In recent years, the parasitic disease which is transmitted

by insects, has spread to parts of Asia and Latin America. Tests show the insect control strategies developed by ILRI can work. In test areas, both income and nutrition improved. "Best of all," notes ILRI Director General Hank Fitzhugh, "assessments of impact from the trypanosomiasis research indicate important improvement to the income and nutrition of the smallholder families without significant environmental problems."

The growing importance of livestock production in the food system, then, plays a double role, for livestock production is profitable for farmers and has substantial impact on poverty reduction. The addition of meat and dairy products enriches the nutrient-deficient diets of both poor urban consumers and their rural counterparts who subsist on starchy staples.

In addition to work to improve animal health, ILRI researchers are using genetic markers to improve resistance to disease and parasites and to improve livestock production. In Ethiopia, the income in households with improved crossbred cows was seven times higher than in those with local cows. In Kenya the difference was even greater.

ILRI researchers have also shown small farmers the advantages of a new technology for plowing with dairy cattle, eliminating the need to keep and feed oxen. Working with researchers from other centers on livestock-related nutrient cycling, they have documented the importance of livestock manure and how it can be better managed to increase crop production and maintain soil fertility.

SAFEGUARDING GENETIC RESOURCES. IPGRI is one of the world's most active collectors of valuable and endangered plant genetic resources. By 1997, IPGRI had collected over 200,000 samples that otherwise might have been lost. Almost 80,000 of these accessions are held in trust in CGIAR genebanks where they serve the scientific needs of the world community.

National and regional genebanks also play a critical role in the conservation of the agricultural biodiversity. With IPGRI's help, valuable plant genetic resources are being safeguarded. By 1996, in addition to the CGIAR genebanks, there were over 1,300 national and regional genebank collections, of which 397 are maintained under long- or medium-term storage conditions.

IPGRI's International Network for the Improvement of Banana and Plantain (INIBAP) has established the world's largest collection of banana and plantain, the developing world's fourth most important crop.

IPGRI helped FAO to prepare the International Technical Conference on Plant Genetic Resources and the resulting *Global Plan of Action*, which was adopted by 150 countries. The plan guides the world's activities in the conservation and use of genetic resources well into the next century.

BUILDING RESEARCH CAPACITY. ISNAR helps developing countries to improve the performance of their agricultural research systems and organizations. It supports institutional development and promotes appropriate policies and funding for agricultural research. During its almost 20-year history, ISNAR has worked with more than 60 developing countries, conducting research and providing training and advisory services to help them improve their agricultural research policies, organizations, and management.

Much of ISNAR's work focuses on strengthening the voice of developing countries in international dialogue related to agricultural research. In its role as a facilitator and continuing supporter of the Global Forum on Agricultural Research, ISNAR works with regional organizations to enhance their participation in setting and prioritizing the research agenda of the CGIAR.

These accomplishments in agriculture research position the CGIAR to confront the emerging global challenges — especially poverty, food security, protecting the environment, and preserving genetic resources.

CGIAR CENTERS

- International Center for Tropical Agriculture (CIAT)
  Cali, Colombia
  Phone: (57-2) 4450000
  Web: http://www.ciat.cgiar.org

- Center for International Forestry Research (CIFOR)
  Bogor, Indonesia
  Phone: (62-251) 622 622 (operator)
  Web: http://www.cgiar.org/cifor

- International Center for the Improvement of Maize
  and Wheat (CIMMYT)
  Mexico City, Mexico
  Phone: 52 5804 2004
  Web: http://www.cimmyt.mx

- International Potato Center (CIP)
  Lima, Peru
  Phone: (51-1) 349-6017
  Fax: (51-1) 349-5638
  Web: http://www.cipotato.cgiar.org

- International Center for Agricultural Research In the
  Dry Areas (ICARDA)
  Aleppo, Syrian Arab Republic
  Phone: (963-21) 2213433
  Web: http://www.cgiar.org/icarda

- International Center for Living Aquatic Resources
  Management (ICLARM)
  Makati City, Philippines
  Phone: (63-2) 812-8641 to 47
  Web: http://www.cgiar.org/iclarm

- International Centre for Research
  in Agroforestry (ICRAF)
  Nairobi, Kenya
  Phone: (254-2) 521450
  Web: http://www.cgiar.org/icraf

- International Crops Research Institute for the Semi-
  Arid Tropics (ICRISAT)
  Patancheru, Andhra Pradesh, India
  Phone: (91-40) 3296161
  Web: http://www.cgiar.org/icrisat

- International Food Policy Research Institute (IFPRI)
  Washington, DC, United States
  Phone: (1-202) 862-5600
  Web: http://www.cgiar.org/ifpri

- International Institute of Tropical Agriculture (IITA)
  Ibadan, Nigeria
  Phone: (234-2) 2412626
  Web: http://www.cgiar.org/iita

- International Livestock Research Institute (ILRI)
  Nairobi, Kenya
  Phone: (254-2) 630743
  Web: http://www.cgiar.org/ilri

- International Plant Genetic Resources Institute
  (IPGRI)
  Rome, Italy
  Phone: (39-06) 518921
  Web: http://www.cgiar.org/ipgri

- International Rice Research Institute (IRRI)
  Los Baños, Philippines
  Phone: (63-2) 8450563
  Web: http://www.cgiar.org/irri

- International Service for National Agricultural
  Research (ISNAR)
  The Hague, The Netherlands
  Phone: (31-70) 3496100
  Web: http://www.cgiar.org/isnar

- International Water Management Institute (IWMI)
  Colombo, Sri Lanka
  Phone: (94-1) 867404
  Web: http://www.cgiar.org/iwmi

- West Africa Rice Development Association (WARDA)
  Bouaké, Côte d'Ivoire
  Phone: (225) 634514
  Web: http://www.cgiar.org/warda

CGIAR Secretariat
The World Bank
1818 H Street NW
Washington, DC 20433 USA
Phone: (202) 473-8951
E-mail: cgiar@cgiar.org *or* cgiar@worldbank.org
Website: http://www.cgiar.org

PHOTOGRAPHY – Cover: IRRI, Ram Cabrera; Asian Development Bank: p. 5; Curt Carnemark/World Bank: pp. 12, 14, 17, 21, 23, 34, 42, 51, 53, 57, 58, 67, 68 Kay Churnush: p. 6; CIMMYT: p. 28 (bottom right); CIP: p. 28 (bottom left); ICARDA: pp. 29 (top), 64; ICRISAT: pp. 29 (middle, right), 55; IITA/Mark Winslow: p. 3; Robb Kendrick: p. 20; David Kinley: pp. 28 (middle), 31, 38, 62; Kay Muldoon: p. 30

ABOUT THE AUTHORS

Mahendra M. Shah

Mahendra Shah has served as Executive Secretary to the CGIAR System Review, based at the World Bank in Washington, since May, 1997. For almost three decades, Mr. Shah has specialized in sustainable development, economic planning, and emergency relief operations. He started his career at the University of Nairobi and the Ministry of Economic Planning in Kenya.

Mr. Shah served as Senior Scientist at the Food and Agriculture Program, International Institute for Applied Systems Analysis and was Director at IIASA of the UNFPA/FAO/IIASA Land Resources for Populations of the Future Study. He was Director of the UN Office for Emergency Operations in Africa, where he was responsible for the information system that mobilized and delivered US$3.5 billion in emergency aid. He also served in the UN Office of the Coordinator for Afghanistan as Director of Information.

Mr. Shah was a Special Advisor to the Secretary General of UNCED. He prepared the Earth Summit report "The Global Partnership for Environment and Development – A Guide to Agenda 21." Mr. Shah's other appointments include advisor to WHO, IUCN, Earth Council, International Development Center of Japan, and the Systema Alimentario Mexicano, Mexico. He has participated in UNFAO, UNICEF, UNDP, and UN missions to over 30 countries in Africa, Asia, and Latin America.

Mr. Shah received his Ph.D. from the University of Cambridge in 1971. He has co-authored a book on development planning in Kenya and has written articles for the *Financial Times, The Guardian, International Herald Tribune,* and *Kenya Nation.* In the private sector, Mr. Shah is also founder and CEO of GreenSpan Designs, which manufactures patented gardening systems for inner city enhancement and for vegetable production.

The Honorable Maurice F. Strong, P.C., O.C., LL.D

Maurice Strong chaired the CGIAR System Review. He is chairman of the Earth Council and also serves as Senior Advisor to both the United Nations and the World Bank. He was recently appointed President of the Council for the United Nations University for Peace in Costa Rica and will also serve as Rector. His other appointments include Director of the World Economic Forum Foundation, Chairman of the Stockholm Environment Institute, and Member and Chairman of the Executive Committee of the United Nations Foundation.

Mr. Strong has held positions as Chairman of the World Resources Institute, Chairman and CEO of Ontario Hydro, as Secretary-General of the United Nations Conference on Environment and Development in 1992 (the Earth Summit), Executive Director of the United Nations Environment Programme, as well as Secretary-General of the United Nations Conference on the Human Environment in 1972.

Mr. Strong is currently Chairman of Technology Development Corporation and Strovest Holdings. He was the first President of the Canadian International Development Agency and also served as Chairman of the Canadian Development Investment Corporation, and President, Chairman, and CEO of Petro Canada.

Mr. Strong's professional accomplishments in Canadian business and international affairs have resulted in a number of honors, including membership in the Queen's Privy Council for Canada, the Swedish Royal Order of the Polar Star, the Order of Canada, the Blue Planet Prize of Asahi Glass Foundation, the Jawaharlal Nehru Award for International Understanding, Commander of the Order of the Golden Ark (Netherlands), Fellow of the Royal Society (UK), the Tyler Prize, the U Thant Prize, and 41 honorary doctorates.